NIGERIAN COALS:

Technical properties and
utilization potentials

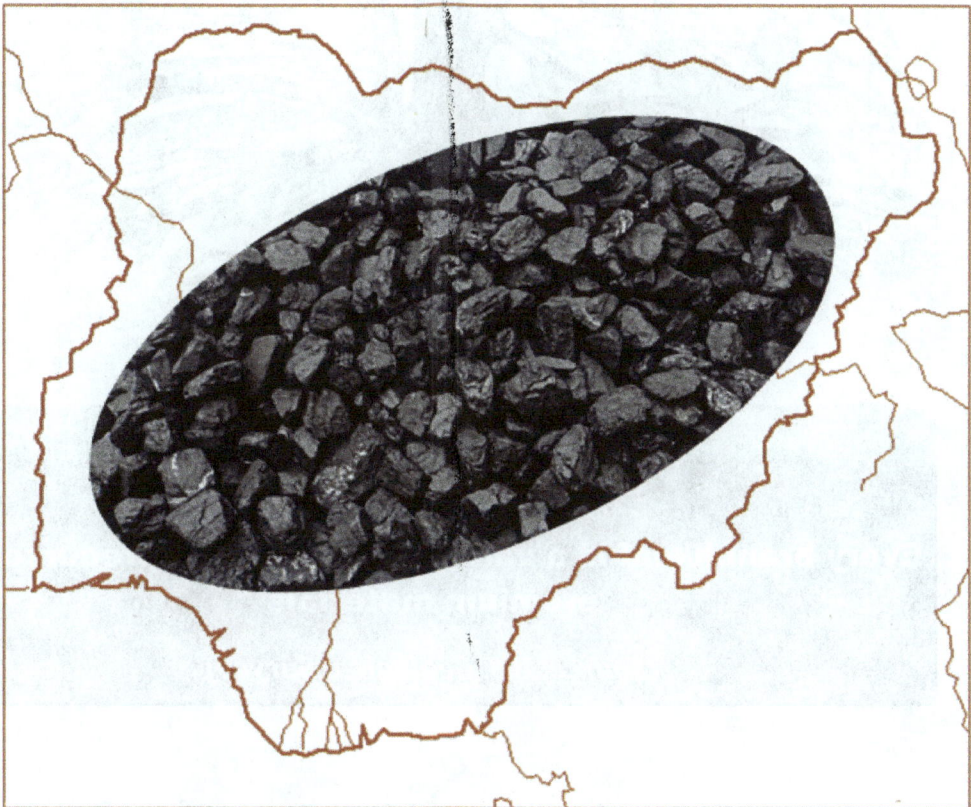

COAL CHEMICALS TREE

Every branch/leaf is a valuable chemical or precursor to premium materials

(Source: U.S. Geological Survey)

Source of cover page IGCC Power Plant
photo: netl.doe.gov

NIGERIAN COALS:

Technical properties and utilization potentials

Adeniyi A. Afonja

SineliBooks

Mansfield, Texas, U.S.A.

ISBN: 978-0-9985843-2-4

Typeset by McConsult, Mansfield, Texas, U.S.A.
Printed in the United States of America
First Impression 2017
Limited edition printed in colour

Contents

Acronyms

ASTM	American Society for Testing and Materials
BS	British Standards
CBM	Coal bed methane
CCGT	Combined cycle gas turbine
CPT	Crossing Point Temperature
CRL	Coal Research Laboratory
CSP	Concentrated solar power
CTL	Coal-to-Liquid
CVD	Chemical vapour deposition
DAF	Dry, ash-free basis
DCL	Direct coal liquefaction
DISCOs	Distribution Companies
DSC	Delta Steel Company
EDI	Energy Development Index
EPSR	Electric Power Sector Reform
F-T	Fischer-Tropsch
HAP	Hazardous air pollutants
HDI	Human Development Index
HGI	Hardgrove Grindability Index
HPHT	High pressure-high temperature
IEA	International Energy Agency
IGCC	Integrated Gasification Combined Cycle
ICL	Indirect coal liquefaction
JISF	Japan Iron and Steel Federation
LNG	Liquefied natural gas
LPPFO	Low pour point fuel oil
M_{10}	Micum Coke Abrasion Index
M_{40}	Micum Coke Strength Index
MEG	Monoethylene glycol
MJ/m^3	Mega Joules per cubic metre
MMSD	Ministry of Mines and Steel Development
Mpa	Mega Pascals
Mtoe	Million tonnes of oil equivalent
NDHC	Niger Delta Power Holding Company
NCC	Nigerian Coal Corporation
NEPA	Nigerian Electric Power Authority
NEPP	National Electric Power Policy
NERC	National Electricity Regulation Commission
NESI	National Electricity Supply Industry
NGIM	National Gas Infrastructure Masterplan
NIOMCO	National Iron Ore Mining Company

NIPP	National Independent Power Project
OCGT	Open cycle gas turbine
OCR	Oxygen consumption rate
OECD	Organization for Economic Cooperation and Development
PHCN	Power Holding Company of Nigeria
ROM	Run of mine
SNG	Synthetic natural gas
SPP	Self Power Providers
TCN	Transmission Company of Nigeria
TPES	Total Primary Energy Supply
UCG	Underground coal gasification
UNECE	United Nations Economic Commission for Europe
UK-NCB	United Kingdom National Coal Board
WEC	World Energy Council

List of figures

Chapter 4

Chapter 5

List of tables

Preface

In 1903, Geologists from the Imperial Institute of England began a comprehensive survey to determine the mineral resources of Nigeria which was then a British Protectorate. Within a few months coal of lignite rank was discovered in Ogwashi-Ukwu. Further survey resulted in the discovery of higher quality coal of sub-bituminous rank in Enugu area in 1909. Mining had begun by 1915 and initial exploitation was for export to fuel British naval ships during the first world war. A railroad was built to link the mines with the port at Port Harcourt to facilitate export and this marked the beginning of the Nigerian railway system which was powered by coal-fired steam locomotives. Subsequently, coal was used for electric power generation, initially to serve the residences of the British expatriates but use was eventually extended to other parts of the country when two power stations were established at Oji river near Enugu and Ijora in Lagos. With increasing demand, Enugu coal was mined at several faces and other deposits in Okaba and Orukpa further north of Enugu were exploited. For the next fifty years coal played a prime role in the national energy mix, supplying over 70% of the national commercial energy requirements. At some point, Nigeria was exporting coal to neighbouring African countries. The decline in coal demand began in 1960 following the discovery of oil in commercial quantities at Oloibiri and the commissioning of the hydroelectric power station at Kainji, and the industry finally collapsed in 1999.

From the 1920s to the 1950s, numerous tests were carried out on Nigerian coals mostly in British laboratories to determine the technical properties and potential utility. The studies included chemical analysis and carbonization tests. More geological surveys located new coal deposits. The first fully equipped coal research laboratory in Nigeria was established at the University of Ife in 1972 and over the following four decades, has developed an extensive database on local and foreign coals. The primary objective of this book is to synthesize into a reference compendium all available data on Nigerian coals, many in the archives of Geological Survey, Nigerian Coal Corporation, federal ministries, and foreign laboratories. My research and consultancy on coal over the last forty years gave me access to all these rich data sources and my hope is that this book (and its companion volumes: Basic Coal Science and Technology, and Fossil Energy Use and the Environment) will be a useful starting point for future researchers on coal in general and Nigerian coals in particular.

I wish to acknowledge with immense gratitude the invaluable support that I received over the years in terms of funding, fellowship awards, and access to equipment and facilities. Top on my list is University of Ife, (Now Obafemi Awolowo University), Ile-Ife, Nigeria for the first grant that enabled me to establish a Coal Research Laboratory. I am also indebted to many institutions at which I carried out a substantial part of my research, notably, Coal Research Establishment, Stoke Orchard and Northern Carbon Research Laboratories, University of Newcastle Upon Tyne, both in the United kingdom, National Research Council Coal Research Establishment, Ottawa, Canada, and Department of Materials Science and Engineering, Massachusetts Institute of Technology, USA. In 1970, I was awarded a United Nations Development Program Fellowship for a six-month in-plant training in iron and steel technology in Soviet Union steel plants. This gave me an invaluable opportunity for hands-on training in every section of iron and steel technology, from ore preparation to finished products and the experience proved invaluable in my subsequent research and consultancy activities.

My visits to leading coal research establishments abroad gave me invaluable access to state-of-the-art analytical facilities and first rate libraries on coal science and technology, and also afforded me the opportunity to interact with the world's leading researchers on the subject. My numerous research and conference trips abroad were funded through grants and fellowships, in particular, University of Ife (now Obafemi Awolowo University) travel grants, Commonwealth Research Fellowship (UK), United Nations Development Programme Fellowship, Senior Commonwealth Research Fellowship (UK), Fulbright Fellowship, (USA), Senior Fulbright Fellowship (USA), and National Research Council Fellowship, (Canada).

Finally, I am indebted to my wife, Professor Simi Afonja whose life inspires me on a daily basis. I dedicate this book to her, not only for her support, keeping the home front stable on my numerous field trips and visits to foreign research establishments for over four decades, but also for her forbearance and encouragement. She also contributed in no small measure to the final manuscript by reading over many times and making very useful suggestions. I also wish to acknowledge the invaluable support of Dr. Akinjide Akinola, one of my former students who worked with me on the coal project in the 1980s. He has been a prime mover and motivator in getting me to finish work on this book and the companion volumes which started many years ago. I also recognize with deep gratitude the contributions of many other former students who worked with me in various phases of the coal research project over four decades.

Adeniyi A. Afonja
Professor Emeritus
July, 2017

1 Coal properties, origin and utilization potentials

1.1 INTRODUCTION

Coal is a combustible black or dark brown sedimentary rock consisting mainly of carbonized plant matter. It belongs to a family of primary fuels known as *fossil* fuels (because they are believed to have derived from dead plants, animals and fossils). The other fossil fuels (oil and gas) are believed to have originated from buried terrestrial matter but they derived mainly from marine animal remains buried beneath oceans. Because they are lighter and fluid, oil and gas can migrate through cracks and, although most of the world's oil and gas resources are believed to be beneath oceans, substantial deposits are often found in dry land due to migration, or ocean recession. In contrast, coal remains in the original location of formation unless moved by ocean intrusions, earth movements, etc.

Coal has been in use as an energy source, probably from the Stone Age when the early man made fire from wood and outcrop coal pieces, lit by the splint method. Evidence abounds that coal was used extensively for smelting copper as early as 3000-2000 BC, probably obtained from outcrops and surface mining. There is also ample archeological evidence that coal was used extensively in the Iron Age for smelting and working iron. By the end of the second century AD small-scale coal mines had been established in England and other European countries and coal trade had started to flourish. The Industrial Revolution of the 17th and 18th centuries was powered by coal, used in driving steam engines and trains, production of iron and steel and other metals, industrial and domestic heating, etc.

Coal is a vital energy source, supplying around 30% of the global primary energy. Dependence on coal for primary energy varies between countries, from less than 30% to over 90%. Coal is also the largest source of energy for electric power generation, accounting for over 40% worldwide. Compared with the other fossil fuels (oil and gas) which are available in a few regions of the world, coal is widely available worldwide and has been found in over one hundred countries. The identified and proven reserves are believed to be less than 40% of the potential global coal resources. Many countries remain unexplored and even current coal producing countries are believed to have much higher coal resources than proven so far.

Coal has been an important primary energy source for thousands of years. It has also been the main source of energy for the extraction of metals and production of cement. Apart from its use for cooking, heating and metal extraction and materials production, coal is an important raw material for the production of many other valuable products, including synthetic gas, hundreds of important chemicals, and high technology materials (graphite, industrial diamond, carbon fibre, carbon nanotubes, graphene, etc).

Coal has been a subject of extensive research for over a century and a lot is known about its origin, physical, chemical and thermal characteristics. While only a summary of the science and technology of coal will be presented in this chapter, detailed information is available in many good books (van Krevelen, 1963; Berkowitz, 1994); Speight, 2012; Thomas, 2013; Afonja, 2017).

1.2   ORIGIN AND FORMATION OF COAL

Coal is believed to have been formed from accumulated prehistoric vegetation that originally accumulated in swamps and peat bogs, eventually buried by earth movements and altered due to high temperature, pressure, and the action of bacteria. All living plants store energy as a result of photosynthesis and when they die and decay this energy is normally released. However, if the deposition conditions are favorable, the process of biochemical decay may be slow or interrupted and most of the energy remains trapped in the matter. The evolution of coal from plant matter is believed to take place in two stages: *peatification* and *carbonification*.

1.2.1   Peatification (Diagenesis)

Plant matter deposited in swamps under favorable environment goes through a sequence of biological, geothermal and geological processes which result in partial decay and conversion to a brown, spongy, woody material known as *peat*. A primary condition for peat formation is that the plant matter is partially submerged in fresh water swamps and peat bogs. The prevailing conditions must be such that the level of bacterial activity is low. If the rate of decomposition of organic material is too rapid, peat will not accumulate. Typically favourable conditions include abundant flora, excess moisture or high water table, low temperature (20-50°C), partially oxidizing conditions and appropriate acidity. The submerged plant material (cellulose, lignin, fats, waxes, resins and proteins) is attacked by aerobic and anaerobic microorganisms. Microbial activity stops when the decaying mass becomes too acidic to support fungi and bacteria (pH of 6.5 to 7.5 facilitates microbial activity, biodegradation ceases at pH of 2 to 3). A typical potential site for peat accumulation (mire) is shown in Figure 1.1. However, it is becoming increasingly unlikely that many potential peat sites will eventually form coal because of the extensive destruction of vegetation by human activities. Furthermore, peat that has already accumulated in mires is being used extensively as fuel and for several other purposes and the chances of eventual burial are slim.

The first stage of *coalification* involving the accumulation of organic matter in a suitable environment for peat formation is known as *peatification* or *diagenesis* and marks the end of the biochemical process. Subsequent changes involve thermo-mechanical processes of pressure and temperature. The peat formed may remain in-situ while sediments are piled on it by earth movement or marine incursion. The subsequent burial of peat due to sequential deposition of sediments induces pressure which sinks the peat to greater depths. The combined effect of increasing pressure and the natural geothermal temperature gradient promotes the process of *carbonification.*

Figure 1.1   A potential site for peat formation (peat pond, wetland, mire). *(sites.google.com).*

1.2.2 Carbonification (Metamorphism)

Metamorphism (carbonification) is the second stage of the coalification process. Pressure, temperature and time are the major factors controlling this step which is associated with a decrease in water content, loss of oxygen and enrichment of carbon which also increases the *calorific value* of coal. Succeeding layers of sediments deposited on the peat due to earth movements, marine incursions, etc subject the peat to increased pressure, with heat provided by the natural geothermal gradient. Carbonification of peat results in the formation of *lignite*, the youngest coal. Over millions of years, lignite matures into *bituminous*, and ultimately, *anthracite* coal (Figure 1.2). Depending on the rank, the carbon content can vary from around 50% to over 95% as the coal matures. Although peat is the precursor of coal and a fuel, it is not normally classified as coal. Lignite is considered as the youngest coal and anthracite the most mature. However, technically, graphite is the highest rank of coal because it is of similar plant origin and has the highest carbon content, but there are significant differences in the geological and geothermal conditions of formation.

Graphite deposits are often found at great depths under hills and rocks and are believed to have been formed under much higher temperatures and pressures compared with coal. Some graphite deposits may have originated from coal deposits on which heavy rock and sediments were deposited as a result of earth movements. Graphite is not usually considered along with coals as fuel because it is difficult to ignite and burn, hence it is not commonly used as fuel. Graphite is mostly used for the manufacture of pencils, high-temperature refractories, high-temperature lubricants, electrodes, radiation shields, high-temperature crucibles, etc. Graphite is also a precursor of many high-technology materials including industrial diamond, carbon fibre, carbon nanotubes, graphene, etc.

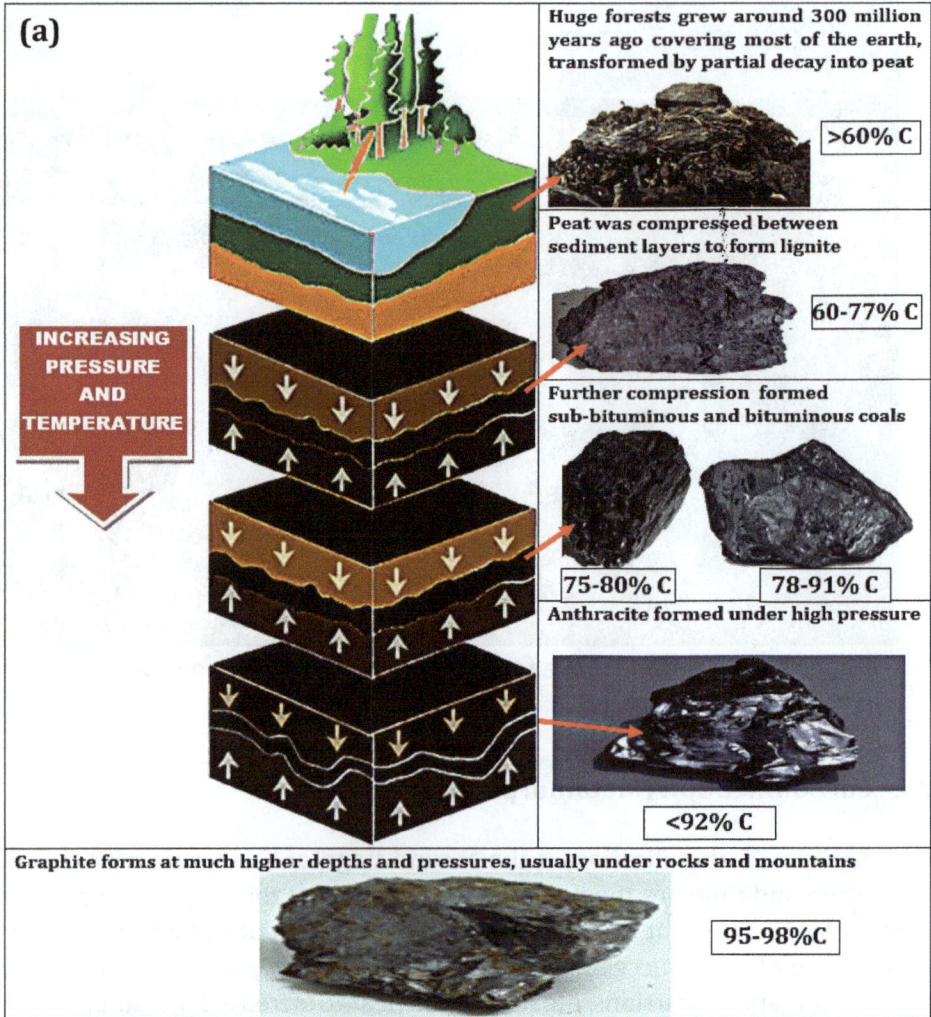

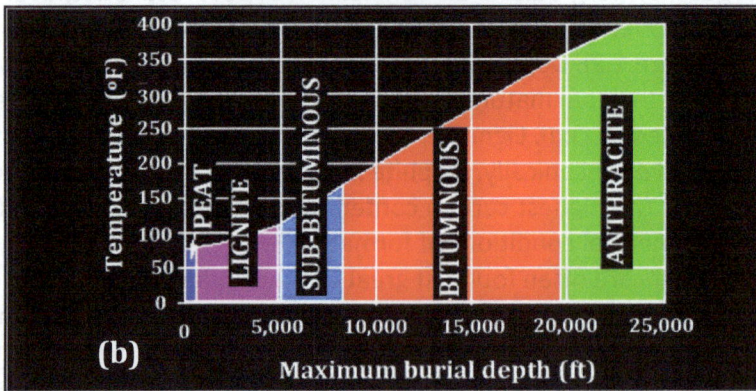

Figure 1-2   (a)   The process of coal formation from plant matter. *(Images from W. H. Freeman & Co., 2007; geology.com; geologycafe.com; indigoplanetimages.com; miningmckissick.wordpress.com).*

(b) Effect of the Earth's natural thermal gradient on coalification *(Adapted from geocraft.com).*

1.2.3 Geological history of coal deposits

There are two theories on subsequent events in the process of coalification after peatification. One theory (the *growth-in-place* or *autochthonous theory)* is that the plants were buried on the same site of peatification and coalification (carbonization) occurred due to deposits of earth and sediments on the peat by earth movements and river and marine transgression. A second theory *(allochthonous theory)* suggests that the peat may have been transported from the original deposition site by flood intrusions. There is ample geological evidence to support both theories. Coals that develop in-situ from peat formed in fresh water tend to contain less sulphur and other mineral matter compared with coal formed from displaced peat deposits.

1.3 COAL PETROGRAPHY

Coal is an inhomogeneous mixture of physically distinctive and chemically different organic constituents known as *macerals* or *microlithotypes*, and inorganic minerals which are classified as mineral matter. The heterogeneity of coal derives from the diversity of source materials which accumulated in the peat swamp, as well as the geological history of the deposit.

1.3.1 Coal macro-petrography

Visual examination of coal often reveals prominent structures (bright and dull bands, charcoal-like structures). Some coals are dull and amorphous while others show embedded bright bands. Lignite and sub-bituminous coals (low-grade coals) have brown colour, woody structure, bituminous coals are black, with banded structure, while anthracite is black and lustrous. Low-grade coals burn with very smoky yellow flame, bituminous coals also burn with yellow flame but are less smoky. Anthracite burns with little or no flame.

1.3.2 Coal micro-petrography

Microscopic examination of coal reveals organic and inorganic constituents of distinctive physical (and chemical) characteristics. The organic micro-constituents are classified into three major groups known as *macerals*: *vitrinite*, *liptinite* and *inertinite,* with many sub-groups within each group (Table 1.1). The inorganic matter is classified as mineral matter. As coalification advances the physical and chemical properties of all coal macerals converge and are virtually indistinguishable in anthracite.

1.3.2.1 *Vitrinite*

Vitrinite macerals are believed to have derived from cellular plant material such as roots, bark, plant stems, and tree barks. The maceral group has a high calorific value (24-28 MJ/kg) and holds a large proportion of the volatile matter in coal (24 - 30%). Vitrinite is usually the most dominant maceral in coal and commonly makes up 50 to 90% of the coal structure. In many coals, vitrinite makes up the matrix in which the various other macerals are dispersed (Figure 1.3). However, in some coals (for example, some Canadian and South African coals), the inertinite macerals dominate the coal structure.

Table 1.1. Microscopic maceral groups and sub-groups of coal organic constituents. (*ICCP, 2001; www.intechopen.com*).

Liptinite Group	Inertinite Group	Humanite Group	Vitrinite Group		
Sporinite	Fusinite		Textinite	Telovitrinite	Telinite
Cutinite	Semifusinite	Detrohumanite	Ulminite	Detrovitrinite	Collotelinite
Resinite	Funginite	Gelohumanite	Attrinite	Gelovitrinite	Vitrodetrinite
Alginite	Secretinite		Densinite		Collodetrinite
Suberinite	Macrinite		Corpohumanite		Corpogelinite
Chlorophyllinite	Micrinite		Gelinite		Gelinite
Fluorinite	Inertodetrinite				
Bituminite					
Exudatinite					
Liptodetrinite					

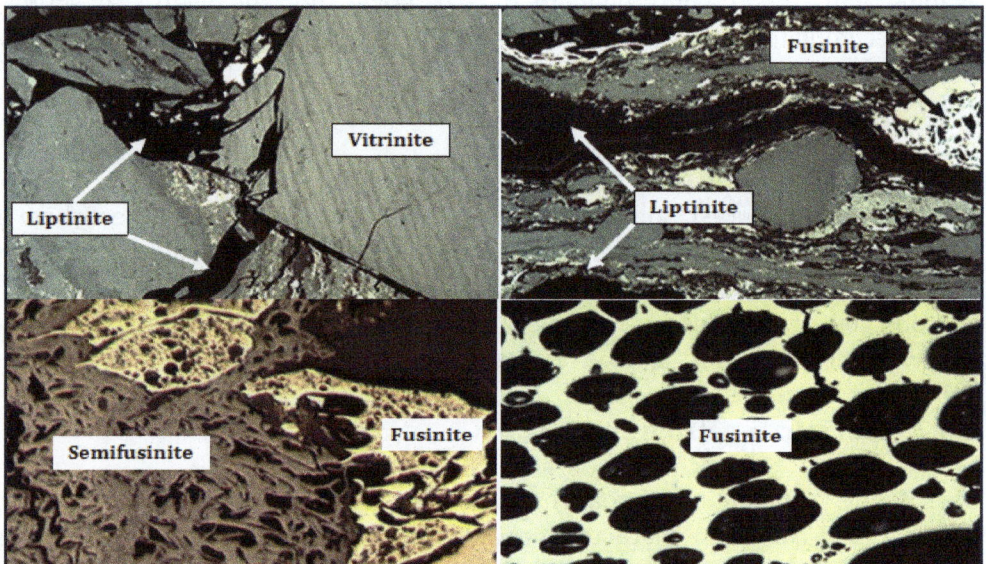

Figure 1.3 The main coal macerals. *(Adapted from siu.edu).*

Vitrinite has a uniform gray colour and it is always anisotropic. The reflectance is in-between those of liptinite (dark, low) and inertinite (bright, high). The morphology of vitrinite in young, low-grade coals may show remnants of plant matter and is classified as *humanite.* This structure disappears as coalfication advances and vitrinite in high rank coals is largely amorphous. The mean maximum reflectance in oil varies from about 0.4% for humanite to up to 5% or higher for vitrinite in high-rank coals. The mean maximum reflectance of vitrinite in oil (Romax) is considered a strong measure of the level of organic maturity of coal (*coal rank*). However, the chemical nature and thermochemical properties of coal constituents are also vital in determining uses for which a particular coal is suitable (*coal type*). Vitrinite in high-rank coals may exhibit a thermoplastic phase depending on the coal type but the humanite group in low-grade coals has little or no thermoplasticity. The thermoplastic properties of vitrinite are determined primarily by its morphology. A coal with largely homogeneous vitrinite will have different thermoplastic properties compared with a coal in which inertinite and/or liptinite is distributed throughout the vitrinite matrix.

1.3.2.2 *Liptinite*

Liptinite macerals are derived from the waxy and resinous parts of plants such as spores, cuticles and resins, leaf matter, spores, pollen and algal matter, which are resistant to weathering and diagenesis. Liptinite macerals tend to retain their original plant form, that is, they resemble plant fossils. Liptinite macerals are hydrogen-rich, thermoplastic, and have the highest calorific values of all coal macerals. Liptinite may be absent in some coals, especially high-rank coals because it tends to disappear as coalification progresses. When present in coal, liptinite may account for no more than 15-20% of the micro-constituents. The maceral group may make up less than 5% of medium-volatile coals and may be absent in low-volatile coals. The outstanding petrographic feature of the liptinite group of macerals is that they all have a reflectance that is lower than the vitrinite macerals in the same coal. Sporinite is the most common of the liptinite macerals and is derived from the waxy coating of fossil spores and pollen.

1.3.2.3 *Inertinite*

The inertinite maceral group has a charcoal-like structure and derived from strongly altered and degraded plant material in the peatification stage of coal formation. Possible causes of thermal degradation are peat fires, volcanoes, etc. Inertinite macerals may occur in coals as small shrads and fragments, or as islands of coherent cell texture (for example, fusinite). Inertinite in some coals (for example, North American coals) may range from less than 5% to 40% whereas the proportion may be higher, up to 70% in other coals (western Canadian, Indian and South African coals). The name "inertinite" derived from the fact that the maceral is thermally inert and does not go through a plastic phase on carbonization. However, inertinite is combustible. The inertinite group has several sub-macerals with different physical and thermochemical properties (see Table 1.1) It is always the highest reflecting maceral compared with the other maceral groups in the same in coal, and fusinite has the highest reflectance of all macerals. The most common type of inertinite maceral is semifusinite and while the other members of the group are thermochemically inert, properties of semifusinite can vary from inert to reactive.

1.3.2.4 *Mineral matter*

Mineral matter in coal is not a maceral but its presence affects the technological properties of coal significantly. A group of inorganic minerals may become associated with coal during the various stages of coalification. These include inorganic constituents of the original plant material (plant ash), as well as inorganic sediments deposited on peat at various stages of coalification. Mineral matter may be intimately intergrown with organic matter in coal. This depends on the conditions in the peat swamp (depth of coverage by water, water movement, chemical composition and acidity of the swamp, etc). Typical intergrown minerals include finely intergrown pyrites and siderite, fine-grained quartz and intercarlations of clay minerals

(Figure 1.4). Some elements in the mineral matter, for example sulphur may be covalently bonded to carbon in the organic structure of coal.

Mineral matter can also be deposited in discrete forms in seam cleats and fissures, the extent depending on activities in the overburden and underburden of the coal seam. Many of these minerals form solutions which may ascend or descend and penetrate into coal seams. The group of mineral matter in coal is important because its presence in coal is significant in determining the economic value and market acceptability of coal. The quantities and physico-chemical characteristics of mineral matter in coal determine whether or not the coal can be cleaned economically. A wide range of mineral matter (over 150 different types have been identified) may be present in coal either intimately intergrown with the coal or as discrete matter deposited in cleats and other fissures in the coal. The principal minerals are clays, carbonates, sulphides, oxides, quartz and salts (chlorides, sulphates, etc). Mineral matter in coal is classified on the basis of its origin into two categories: *inherent* mineral matter and *adventitious* mineral matter. Inherent mineral matter derived from the organic and inorganic constituents that previously formed part of the tissues in the peatification stage. Also, mineral matter deposited in the mire (peat pond) by earth movement or ocean incursion could become inherent in the coal matrix. Most of the mineral matter is chemically and colloidally combined with organic coal matter. It is not removable by washing/cleaning, only by thermal or chemical degradation of the coal aromatic structure.

Adventitious mineral matter derived from post-peatification deposition of mineral matter by oceanic transgression, land slides, etc. These events may occur during or subsequent to coalification. Depending on the geological history of the deposit, adventitious mineral matter may be present in chemical or colloidal combination with the coal substance. Adventitious mineral matter may be in discrete form but finely disseminated throughout the coal structure, or it may be present in coal seam cracks and fissures, seam intrusions, seam overburden, etc.

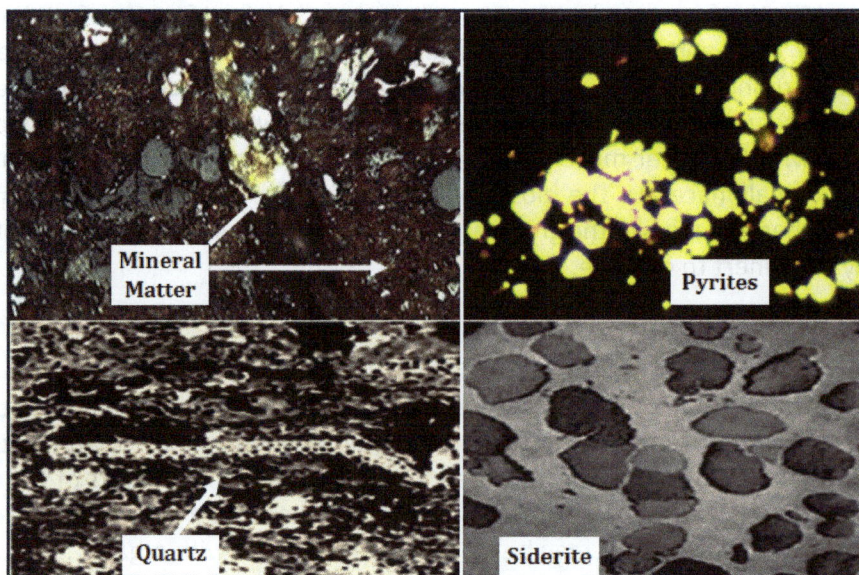

Figure 1.4 Organic and inorganic constituents of coal (macerals and mineral matter). *(Images from (ICCP, 1963; http://coalandcarbonatlas.siu.edu/coal).*

While lump mineral matter can be easily removed from coal by cleaning, finely disseminated mineral matter is difficult to remove and requires that the coal is finely ground. This increases cleaning costs significantly. In addition to the major elements in coal mineral matter, a number of elements are often present in very small amounts and may be present as either inherent or adventitious matter, or in both forms. This group of elements known as trace elements includes mercury, zirconium, zinc, uranium, tin, vanadium, lead, molybdenum, nickel, silver, strontium, germanium, copper, cobalt, bismuth chromium, gallium, beryllium, gallium, antimony.

Mineral matter is not combustible and ends up as ash at the bottom of combustion chambers or in coke. The chemical composition of mineral matter is important in estimating loss of heat value of coal due to endothermic decomposition of mineral matter, and its effect on the fusibility and removal of the ash formed during coal carbonization or combustion. Also, ash chemical characteristics largely determine the type of fluxes and refractory lining required in metal melting furnaces.

1.4 TECHNICAL PROPERTIES OF COAL

As a result of the coalification process, the physical and chemical properties of coal change. Carbon content and calorific values increase while moisture, volatile matter and oxygen decrease as coalification progresses. (Table 1.2). For example, moisture in lignite may be as high as 60% and carbon content as low as 30%, rising to over 90% in anthracite.

1.5 COAL RANK, TYPE AND CLASSIFICATION

The degree of coalification (coal *rank*) and nature and relative proportions of the macerals (coal *type*) largely determine the technological properties of coals and the end use for which they are suitable.

Table 1.2 General range of major elements in coal and vitrinite reflectance (Romax) as a function of rank.

Coal (% dmmf)	Carbon (% dmmf)	Hydrogen (% dmmf)	Oxygen (% dmmf)	Volatile matter (% dmmf)	Calorific Value (Mj/kg dmmf)	Inherent moisture (% wet basis)	Romax (%)
Peat	50-55	8-10	45-55	75-55	17-19	40-30	
Lignite	60-77	6.4-4.9	33-15	65-38	22-30	20-15	< 0.27-0.38
Sub-bituminous	75-80	5.6-5.1	17-12	47-38	29-33	16-12	0.38-0.67
Bituminous	78-92	5.8-4.4	15-2.5	47-19.	33-35	15-0.4	0.47-2.1
Semi-anthracite	91-94	4.8-3.7	3.5-1.9	19-9	34-36	0.5-2.0	2.1-3.5
Anthracite	92-96	3.9-1.9	2.8-1.8	9-2	34-37	0.9-2.9	> 3-3.5
Graphite	96-99	0	0		35-37		

1.5.1 Coal rank

As discussed in an earlier section, coal is ranked primarily on the basis of the degree of coalification as indicated by various parameters, including vitrinite reflectance, chemical composition and heating value. The main ranks of coal are lignite, bituminous and anthracite coal. However, many coal deposits are in transition between two ranks, hence the sub-divisions of sub-bituminous and semi-anthracite coals. Typical technical properties of the different coal ranks are shown in Table 1.2 and Figures 1.5 & 1.6.

1.5.2 Coal type

Coal type is determined by the nature, chemical composition, relative proportions and morphology of the constituent macerals of coal. While coal rank indicates the degree of maturity, coal type gives more valuable information on the potential technical properties of coal than rank. For example, the morphology and chemical composition of vitrinite and the relative proportions of the three coal macerals and mineral matter all affect the thermal behaviour of coal. Two coals of the same rank, with similar chemical composition can have very different thermal behaviour (Table 1.3).

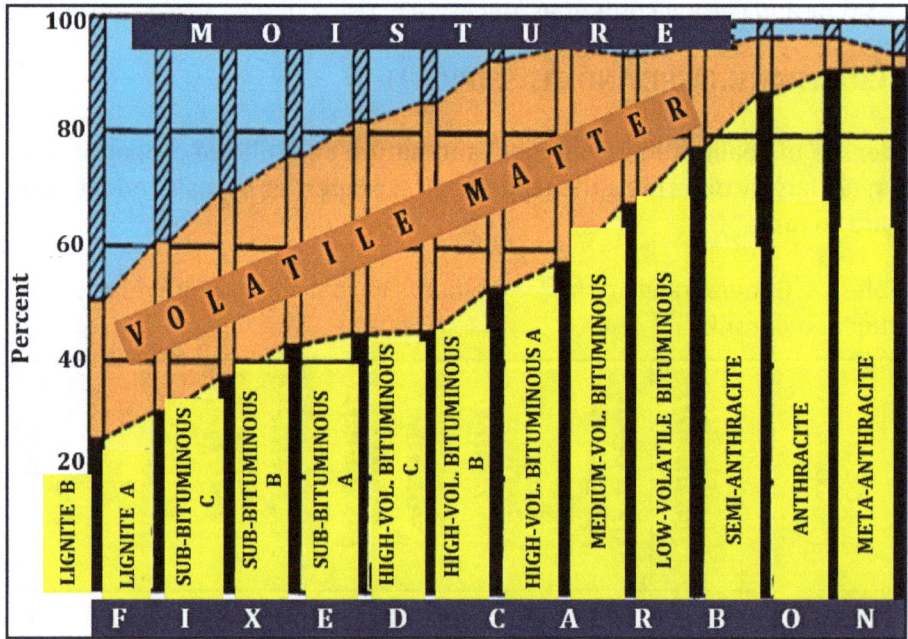

Figure 1.5 The main coal ranks. *(Averitt, 1975).*

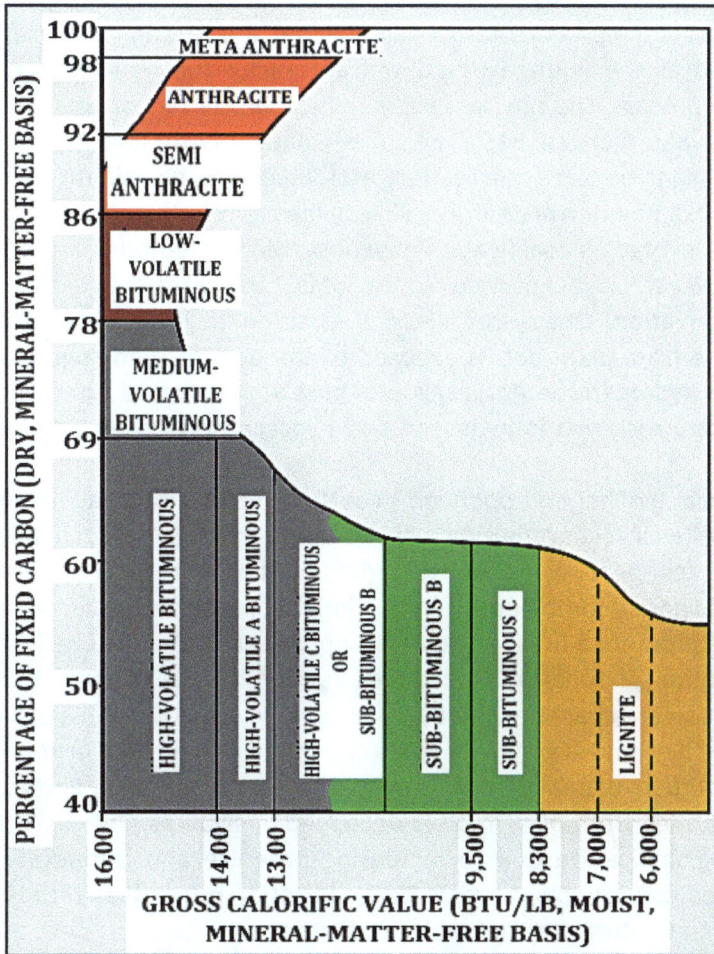

Figure 1.6 Diagram showing classification of US coals by rank. *(pubs.usgs.gov).*

Table 1.3 Thermoplastic of two coals of the same rank and similar chemical composition.

TECHNICAL PROPERTY	COAL A	COAL B
	Rank	
	Bituminous	Bituminous
Inherent moisture (%)	2.5	2.6
Volatile matter (% air-dried)	24.0	24.5
Ash content (% air-dried)	10.0	9.5
Fixed carbon (%)	63.5	63.4
Free Swelling Index (FSI)	9	1.5

1.5.3   Coal classification

Although peat is the precursor of coal and a fuel, it is not normally classified as coal. Lignite is considered as the youngest coal and anthracite the most mature. As a result of the coalification process, the physical and chemical properties of coal change. Carbon content and calorific value increase while moisture, volatile matter and oxygen decrease (Table 1.2). Remains of plant matter (cells, pores, etc) which are prominent in low rank coals under microscopic examination are hardly visible in high rank anthracite. Also, physical and chemical properties converge as coalification advances. Many methods of classifying coal have been developed over the last century or so, the oldest being the simple classification on the basis of visual observation. One system classifies coals on the basis of their source of genesis. Coals that originate from plant debris exposed to the peatification stage are classified as *humic* banded, (low hydrogen content) coals and those which derived from sapropels (loose deposits of sedimentary rock rich in hydrogen and hydrocarbons) are classified as *sapropelic* coals (Figure 1.7).

Sapropelic (*cannel* and *boghead*) coals are believed to have been formed from plant matter deposited in lakes rather than peat swamps (*mires*). Although they may have originated from normal coal-peat swamp matter, addition of other remains, such as algae and wind or water-borne spores, combined with the relative serenity and anaerobic conditions that often exist in lakes promote a breakdown of the organic structure into very fine grains through the biological action of fungi and bacteria (*saprophytes*), a process known as *putrification*. Sapropelic coals are characterized by a non-banded structure and a dull black and sometimes waxy luster. They have high volatile matter content and are rich in liptinites (the microscopic organic constituent of coal derived from waxy and resinous plant parts). Humic coals evolved from peat formed from plant matter deposited in swamps. The projenitor pitch undergoes progressive coalification due to increasing pressure and temperature generated by an increasing sedimentary overburden. Carbonification progresses with time through lignite to bituminous and anthracite coal ranks. Coals that are non-sapropelic are classified as humic coals and constitute the vast majority of coal deposits worldwide.

(a) Humic (banded) coal   (b) Cannel (sapropelic) coal

Figure 1.7 Humic and sapropelic coals   *((a) www.geology.com; (b) www.britanica.com).*

Cannel coals are rich in finely-comminuted vegetable matter (spores, algae and fungal material), and resin globules while boghead coals are rich in algae. During the 19th century, cannel coal was processed by distillation to obtain coal gas and kerosene used extensively for lighting, cooking and heating. Sapropelic coals are thin and lenticular, and rarely occur as large, coherent deposits. They often occur as seam intrusions of limited lateral extent, mostly at the bases and tops of humic coal seams.

Another visual classification system is based on coal rank and end use and the system is used extensively in coal trade (Figure 1.8). Generally, coals are classified into soft, low-rank coals (lignite and sub-bituminous coals), and hard, high rank coals (bituminous and anthracite coals). In the German system, lignite and sub-bituminous coals are classified as *brown coals*. While soft coals are used predominantly for heating, steam raising and electric power generation, hard coals are used for metallurgical extraction and power generation. Most of the international trading in coals involve hard coals.

The Stopes system developed in 1919 classifies coals according to their macroscopic constituents known as *lithotypes*. Ordinary black coal is classified as *clarain*, glossy black coal as *vitrain*, dull rough coal as *durain*, and soft powdery charcoal-like coal as *fusain*. Other classification systems use physical properties (reflectance of the microscopic constituents of coal) or chemical and thermochemical properties (carbon:oxygen or Hydrogen:oxygen ratios, volatile matter content, calorific value, thermoplastic properties, etc) (see van Krevelen, 1963; Thomas, 2013; Speight, 2012; Afonja, 2017).

The earliest published coal classification system developed by Regnault in 1837 (proximate analysis) was based on the determination of the main contents of coal as percentage of total contents (moisture, volatile matter, fixed carbon and ash). This method is still very much in use and is considered in conjunction with calorific value determination as the most basic coal analysis for the evaluation of any coal deposit.

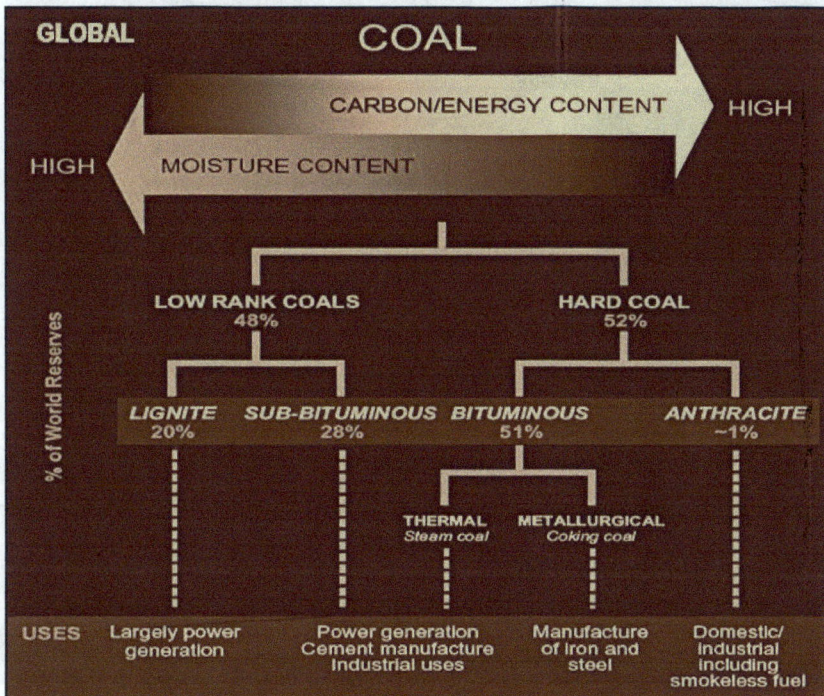

Figure 1.8 Coal classification by end use. *(worldcoal.org)*.

The Seyler system developed at the beginning of the twentieth century was based on six major coal chemical and thermochemical variables: carbon, hydrogen, oxygen, moisture, volatile matter contents, and calorific value. From the 1930s, several other classification systems have been developed, the most prominent being the National Coal Board (UK) and the American Society for Testing and Materials (ASTM-D388) systems, all of which use different parameters for classification, making it difficult to compare coals under the different systems.

The need for a unified coal classification system has led to a more recent classification system (international Coal Classification System) developed by the United Nations Economic Commission for Europe (UNECE) which has been adopted by the International Standards Organization. The system is particularly suitable for classifying coals for industrial applications because it is based on the relevant physico-chemical and thermo-chemical properties of coal (Figure 1.9). Classification according to type involves the relative proportions of both the organic and inorganic constituents. Rank classification is based only on the organic constituents because they are the only substances in coal that are altered by metamorphic processes. The two ranking systems are independent but both should be taken into account in assessing the potential use of coal. As mentioned earlier, two coals of the same rank and similar chemical composition can show very different thermochemical behaviour, one being highly thermoplastic while the other is totally inert when carbonized (Table 1.3).

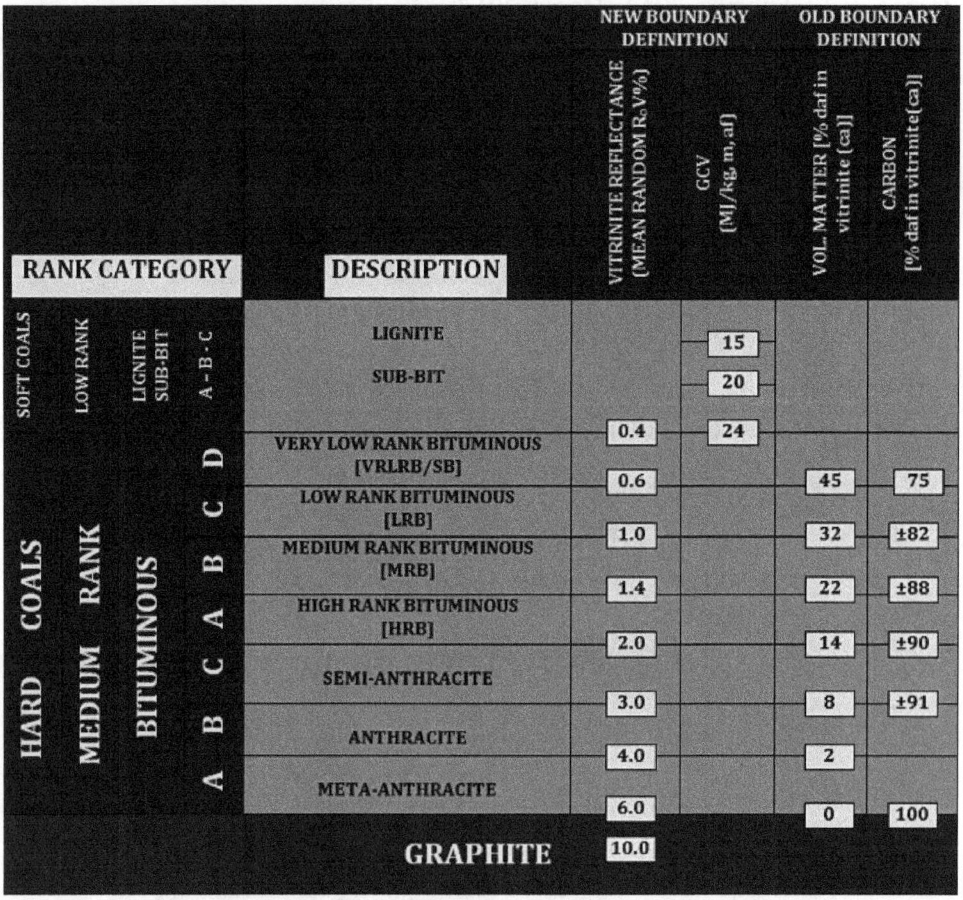

RANK CATEGORY	DESCRIPTION	NEW BOUNDARY DEFINITION		OLD BOUNDARY DEFINITION	
		VITRINITE REFLECTANCE [MEAN RANDOM R_o V%]	GCV [MJ/kg, m, af]	VOL. MATTER [% daf in vitrinite (ca)]	CARBON [% daf in vitrinite(ca)]
SOFT COALS / LOW RANK / LIGNITE SUB-BIT / A–B–C	LIGNITE		15		
	SUB-BIT		20		
HARD COALS / MEDIUM RANK / BITUMINOUS / D	VERY LOW RANK BITUMINOUS [VRLRB/SB]	0.4	24		
	LOW RANK BITUMINOUS [LRB]	0.6		45	75
C	MEDIUM RANK BITUMINOUS [MRB]	1.0		32	±82
B	HIGH RANK BITUMINOUS [HRB]	1.4		22	±88
A	SEMI-ANTHRACITE	2.0		14	±90
C	ANTHRACITE	3.0		8	±91
B	META-ANTHRACITE	4.0		2	
A	GRAPHITE	6.0		0	100
		10.0			

Figure 1.9 New International Coal Classification system. *(www.unece.org)*.

As discussed earlier, there is no universal coal classification system and different terminologies are used for classification in different countries, and also depending on end use. Some of the most common are presented in Table 1.4. It should be noted that there are no internationally accepted limit values to demarcate the various coal groups. Variables used for classification also differ. While some systems use calorific value, some use thermoplastic values, while others use both. The commercial system uses end-use as the primary classification variable. Also, there are classification differences within each system and there are significant overlaps between limits of different systems. For example, some coals which would be classified as lignite in the English system are classified as sub-bituminous in the American system.

Table 1.4 Different sub-divisions and classifications of coals by rank. (IEA, 2013a).

Sub-divisions and Classifications	Increasing coal rank →			
International conventional classification	Lignite	Hard coal		
German system	Brown coal		Hard coal	Anthracite
English system	Lignite	Sub-bituminous coal	Bituminous coal	Anthracite
International system (UNECE)	Lignite	Sub-bituminous coal	Bituminous coal	Anthracite
Commercial system			Steam coal / Coking coal / PCI coal	Steam coal / Anthracite / PCI coal

1.6   GLOBAL COAL RESOURCES, RESERVES AND PRODUCTION

Coal is the most abundant of fossil fuels and the most widely distributed globally. While deposits of oil and gas have been found in relatively few countries, coal has been found on all continents and in many countries around the world. Exploration of coal resources has not been on any scale comparable to prospecting for oil and gas and it is widely believed that the global proven coal resources represent only a small fraction of the potential resources.

1.6.1 Global coal resources and reserves

Global primary energy resources are commonly assessed at three different levels: total resources, total proven reserves, and total recoverable reserves. Resources are inferred from geological and geophysical data, proven reserves are the resources that have been identified, evaluated and quantified with a fair degree of accuracy through borehole drilling, while the recoverable reserves refer to the proportion of the proven reserves that can be recovered economically, using currently available technologies. The limited scope of exploration worldwide and limitations in currently available exploration technologies imply that vast resources probably remain unidentified or proven. Concurrently, there are severe limitations in exploitation technologies. It is anticipated that increased exploration particularly in emerging countries, and future developments in underground gasification, seismic and remote sensing technologies will result in continuous increase in resources and proven reserves of coal and other fossil fuel resources.

Various estimates indicate that less than 50% of the worlds's proven reserves of coal, oil and gas can be recovered economically using currently available exploitation technologies. Large oil and gas deposits are believed to be located at great depths beneath oceans, much too deep and unexploitable by current technologies. Many coal reserves are at over a thousand metres deep whereas current mining technologies can only recover deposits not more than around 500 metres deep. Furthermore, some coal seams are too thin or too sporadic, or the geological and geophysical setting of the deposit are unfavourable for normal mining techniques. Development of new exploitation technologies will inevitably lead to a continuous increase in exploitable reserves. Already, recently available fracturing and well pressurization technologies and successes in shale gas development have led to a very significant increase in the production of oil and gas globally.

Also, developments in mechanized and automated coal mining are making it possible to recover previously unexploitable coal deposits. Furthermore, development of in-situ coal gasification technologies now makes it possible to exploit deposits located too deep for shaft mining or too thin for economic exploitation by existing mining techniques. Previously abandoned coal mines are also being reopened to recover coal pillars by retreat mining or in-situ gasification.

The global coal resources are estimated at 22,000 gigatonnes of which less than 5% is proven and recoverable under current economic and technological conditions (Figure 1.10). Coal exploration is not as economically attractive as oil and gas and reserves are declining while exploitation rate is rising. The view that coal reserves are sufficient to meet global demand in the foreseeable future, the discovery of large unconventional gas resources and a focus on natural gas as a substitute for coal in many industrialized countries have reduced interest in new surveys and new mine developments in the last decade or two (IEA, Resources to reserves, 2013). Proven coal reserves are located in around a hundred countries and many more have potentially large but unexplored reserves. The increasing demand for coal, particularly in the emerging countries, developments in coal gasification, liquefaction, co-power generation and clean coal technologies will inevitably re-energize interest in coal exploration and mine development in the coming years. The bulk of the global reserves of coal is located in four continents and only four countries hold about 75% of the total world reserves of coal, with the United States alone holding about 28% (Figure 1.11). The balance is held by Russian Federation, China, Australia and India (Figure 1.12). About 48% of the total global reserves is of the bituminous, 29% sub-bituminous, and 23% lignite rank. It is clear that China, the USSR and the United States all have dominant position relative to coal resources, both inferred and recoverable.

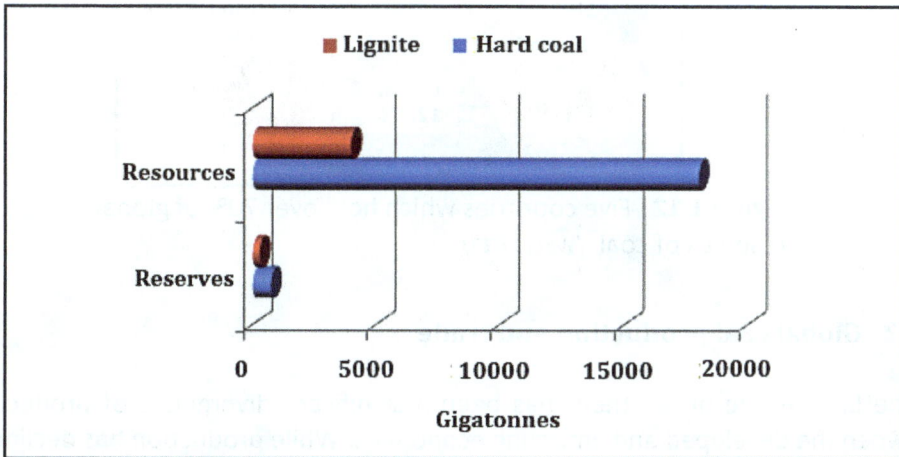

Figure 1.10 World coal resources and reserves. *(IEA, 2013b).*

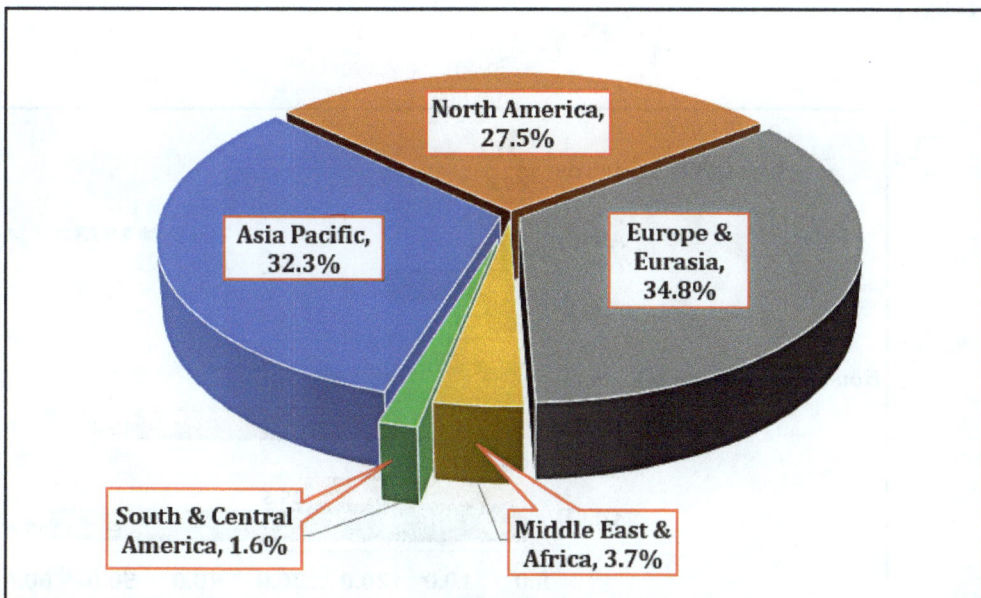

Figure 1.11 World's reserves of coal by region in 2013. (860,938 million tonnes) *(BP, 2014).*

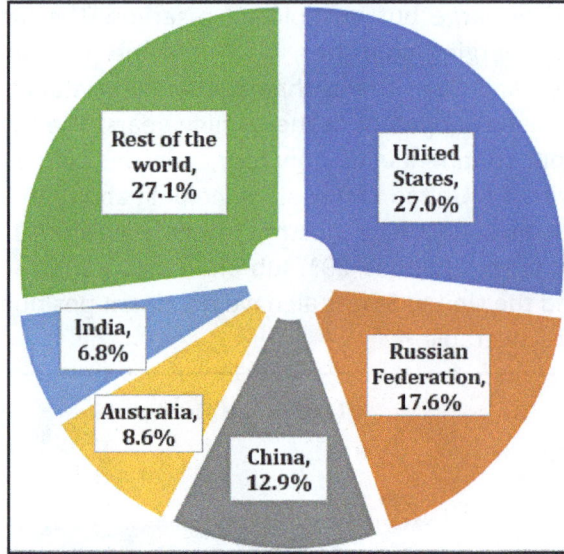

Figure 1.12   Five countries which hold over 70% of global reserves of coal. *(WEC, 2013).*

1.6.2   Global coal production and trade

In the last decade or so, there has been a significant divergence of production patterns between the developed and emerging economies. While production has declined steadily in Europe, output from non-OECD countries has doubled over the same period (Figure 1.13). China is both the largest producer and importer of coal while Indonesia is the leading exporter (Table 1.5).

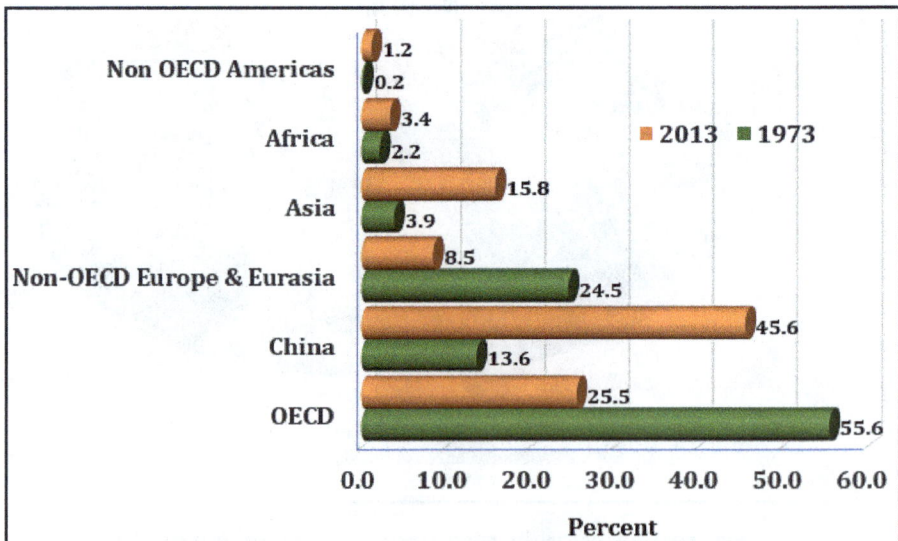

Figure 1.13   Regional share of global coal production in 2013. *(IEA, 2014).*

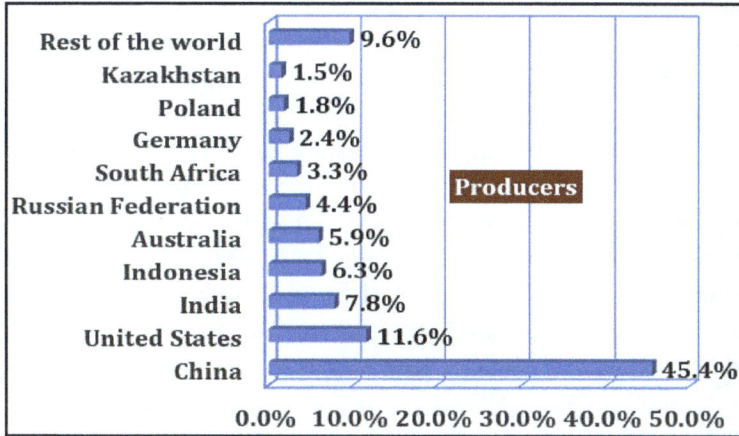

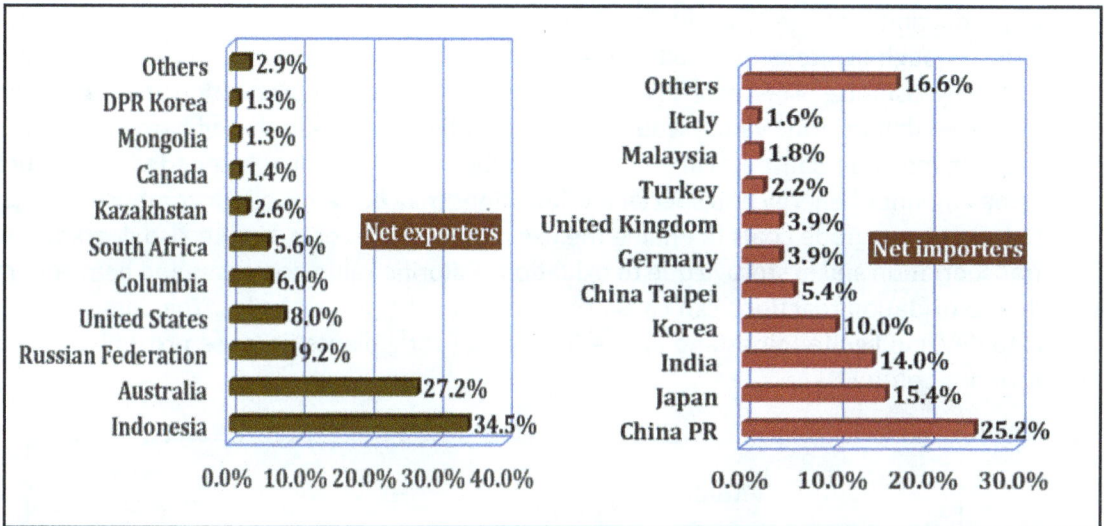

Figure 1.14 Top ten producers, net exporters and net importers of coal in 2013. *(iea.org, 2015).*

In spite of the fact that China, one of the highest coal reserves in the world and one of the leading producers, the country accounted for 23% of the total global imports of coal in 2013 due to increasing local demand, and the cost of domestic production and transportation from the coal deposits in the northern parts of the country to the south-eastern provinces where coal is needed.

1.6.3 Global coal utilization

The dynamics of primary energy utilization have changed significantly over the last four decades, with the share of industrialized countries dropping by nearly 40% (Figure 1.14). The pattern of electric power generation has also shifted. In 1973, OECD countries generated over 70% of global electric power compared with 45% in 2014 (Figure 1.15). Coal supplied nearly 30% of the global primary energy requirements in 2014 and its share has been increasing over the years at the expense of oil (Figure 1.16). However coal share is much lower in the industrialized countries, accounting for less than 20% in industrialized (OECD) countries.

There has also been a significant shift in the pattern of coal production by region in the last four decades. In the 1970s the bulk of global coal supply came from the OECD countries but the pattern has changed significantly, with China alone producing over 45% of the total global output in 2015 (Figure 1.17). Whereas the focus has shifted from coal to gas in the industrialized countries, emerging countries are using more coal for power generation and production of industrial chemicals and high-technology materials. Coal accounted for about 40% of the global electric power generation in 2014 (Figure 1.18). China had the highest power output in 2014, increasing from under 3% in 1973 to 24% in 2014, compared with 45% by OECD countries. China also accounted for around 40% of coal used globally for power generation in 2014, with the United States and India coming far second and third, accounting for 15% and 5% respectively. About 15% was used for steel production and the balance was used for production of chemicals, steam raising, domestic and industrial heating. Only about 15% of the coal produced globally in 2013 was traded internationally, implying that the bulk of production was used locally. International trade in coal is mostly restricted to hard coal and, although lignite is used extensively for power generation, it is very rarely transported over long distances. The calorific value of lignite is low and the production and transportation costs (per unit calorific value) would be significantly higher than that of hard coal because of the high moisture content. Furthermore, a higher tonnage would be needed to produce the same amount of energy and special modifications may be required for combustors designed to burn higher grade coals to enable the use of low-grade coals. Lignite also degrades during transportation and in storage due to oxidation. Calorific value decreases and heat generation due to oxidation reactions can cause self-ignition in lignite coal piles. The common practice is to develop lignite mines and lignite-fired generating plants in close proximity as a single economic entity.

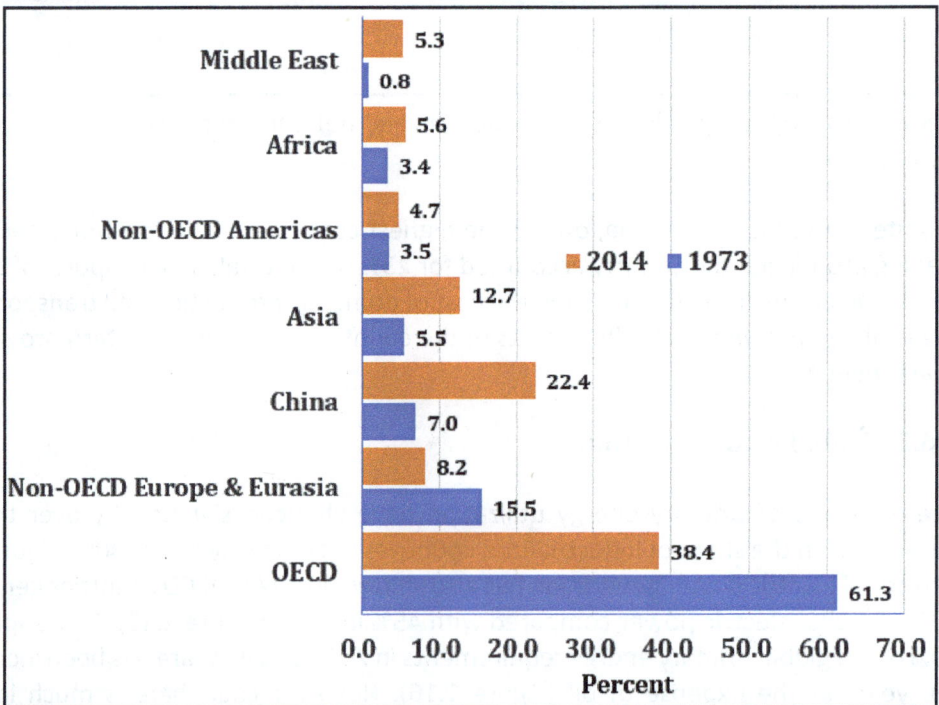

Figure 1.15 Regional shares of global total primary energy supply in 2014. *(IEA, 2016).*

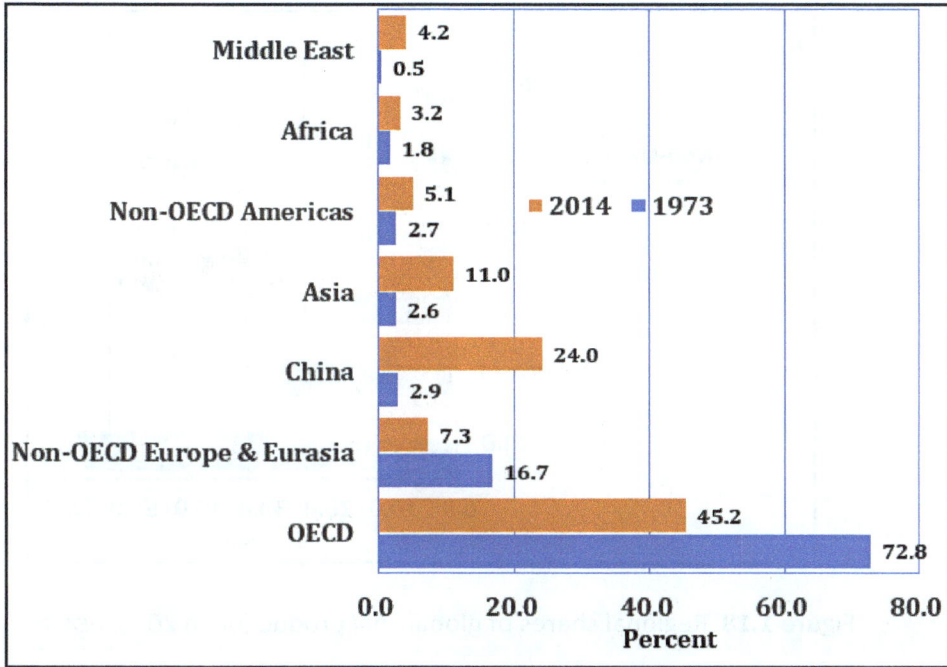

Figure 1.16 Regional shares of global electric power generation in 2014. *(IEA, 2016).*

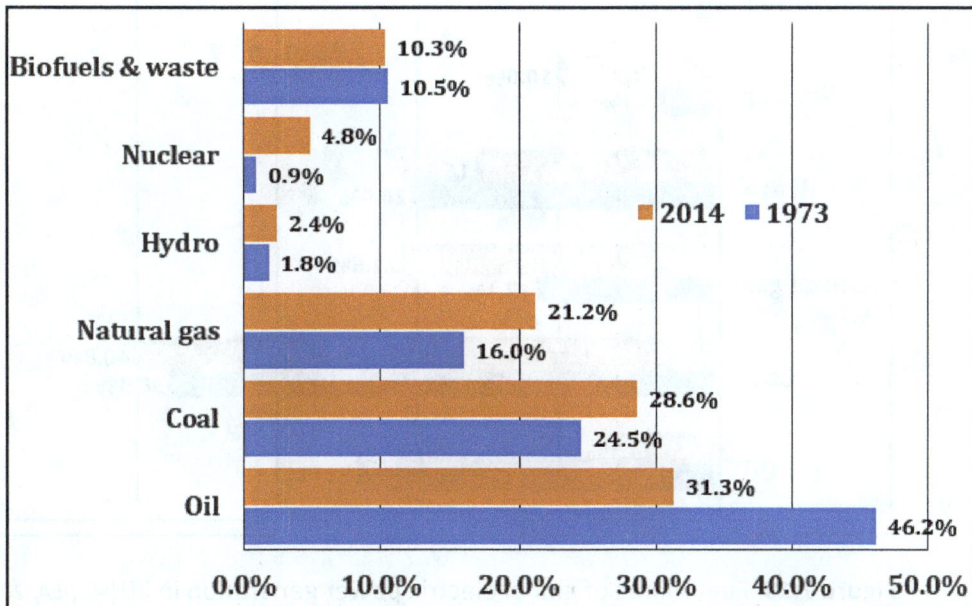

Figure 1.17 Fuel shares of global total primary energy supply in 2014. *(IEA, 2016).*

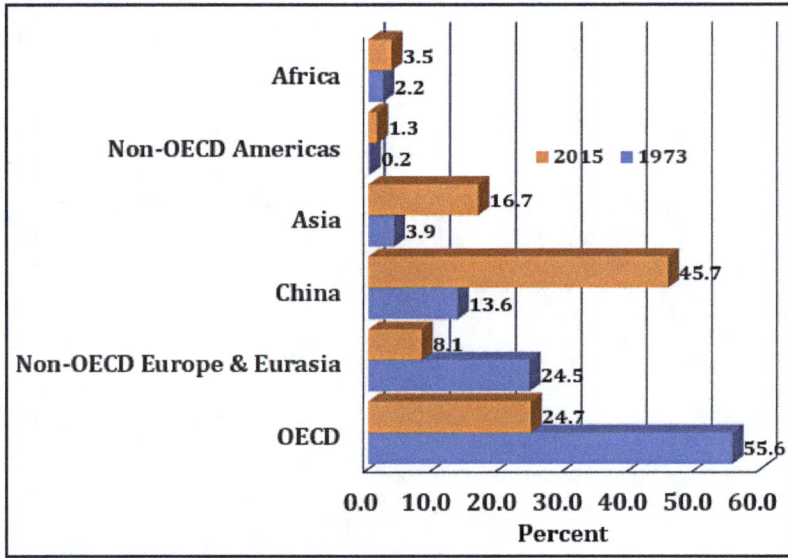

Figure 1.18 Regional shares of global coal production in 2015. *(IEA, 2016).*

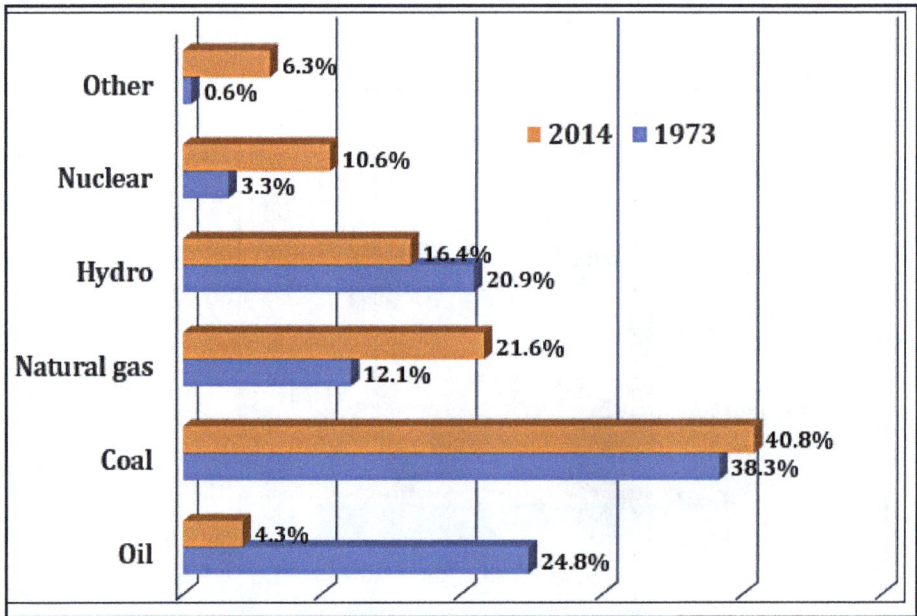

Figure 1.19 Fuel shares of global electric power generation in 2014. *(IEA, 2016).*

1.7 COAL MINING AND BENEFICIATION

A coal deposit is mined if there is enough of it and it is of adequate quality to make it economically recoverable. Other considerations include depth of deposit, seam thickness and continuity, and the nature of the overburden. The depth of coal deposits may range from less than 20 metres to over a thousand metres. Most mines adopt one or both of two mining technologies: opencast and underground mining. The choice largely depends on the geological setting of the deposit.

1.7.1 Surface (opencast) mining

Deposits up to about 50 metres deep are usually mined by opencast techniques. The overburden (earth and rock) above the coal is removed with heavy earth moving equipment (Figure 1.19). Once exposed the coal is drilled, fractured and loaded onto conveyors or trucks. Coal mined by opencast techniques is usually clean with relatively low inorganic content, hence it is often used without cleaning.

1.7.2 Underground longwall mining

Underground mining methods are used when coal is located too deep for opencast mining and one or both of two different mining techniques may be used: longwall mining and room-and-pillar mining. Longwall mining involves using mechanical equipment to open access to medium to thick coal seams. Continuous, rotating mechanical shearers advancing across the face of the coal seam are then used to extract the coal (Figure 1.20). Temporary hydraulic supports are used to hold up the roof of the mine.

Figure 1.20 An opencast coal mine. *(wikimapia.org).*

Chain conveyors are used to move the loosened coal to the surface. Once extraction is completed the roof is allowed to collapse. This method of mining is very efficient and up to 75% of the deposit can be extracted. Furthermore, it is relatively safe since human involvement is minimized. Modern mines use automated, self-advancing hydraulic ceiling supports which travel forward horizontally with the other machinery as the longwall sections are mined, allowing the roof to collapse and fill the void behind them. This system greatly enhances both productivity and safely.

1.7.3   Underground room-and-pillar mining

Room-and-pillar mining method is adopted when the coal seam is small and thin. A network of underground rooms is cut into the coal seam, leaving a series of pillars of coal to support the roof of the mine. Complimentary hydraulic and steel structure roof supports are often installed as well (Figure 1.21). The coal is fractured manually using hydraulic and pneumatic hammers or, in mechanized mines, by using continuous mining machines to cut the coal. Shuttle cars transport the coal to a conveyor belt for further transportation to the surface. This method of mining is inefficient since the coal pillars which may constitute up to 60% of the deposit are left behind. Furthermore, it is relatively unsafe especially if manual fracturing techniques (very common in developing countries) are adopted. Closed room and pillar mines are now being reopened to recover the coal pillars, by a retreating method in which each pillar is fractured and recovered, starting from the farthest and collapsing other pillars in sequence as mining retreats. Some coal pillars are also being recovered by in-situ gasification.

Figure 1.21 Longwall coal mining. *(https://www.victaulic.com).*

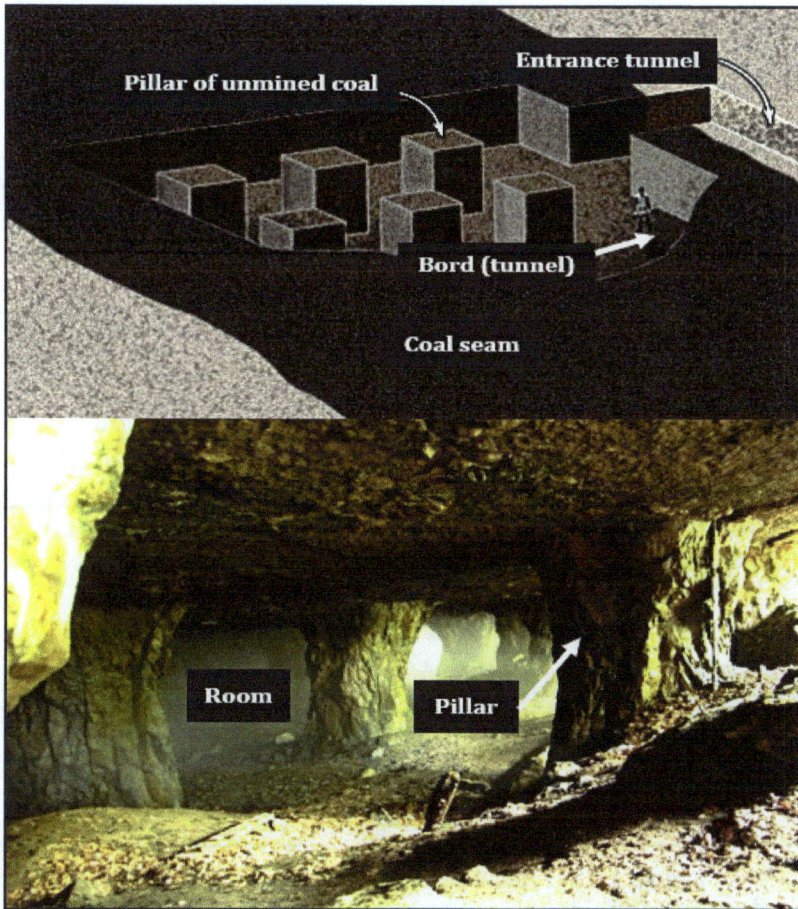

Figure 1.22 Room-And-Pillar mining. *(.teara.govt.nz; kampsenergy.blogspot.com).*

1.7.4 Coal preparation and cleaning

In surface mining, it is usually easy to distinguish between coal and inorganic shale and rock hence coal recovered by opencast mining is usually clean and can be used without further cleaning. However, grinding and sizing may be necessary to meet the specification require-ments of consumers, especially if the coal is to be used in pulverized firing. Coal mined by underground techniques usually contains considerable amounts of rock, shale and lump mineral matter, especially when mining is mechanized. Although some consumers use such coals without cleaning, the heating value is degraded and most customers require that the coal be cleaned. Coal cleaning involves sorting, coarse and fine grinding and separation processes. Inorganic matter is separated from coarse coal by float-and-sink techniques involving washing in solutions of controlled specific gravity. Inorganic matter which has higher density than coal will sink while clean coal floats and is recovered. Fine coal is cleaned by cyclone techniques while ultrafine portions are cleaned by froth flotation. A typical coal cleaning flow diagram is shown in Figure 1-22. Coal cleaning is well treated in many books (see Speight, 2012; Thomas, 2013; Afonja, 2017).

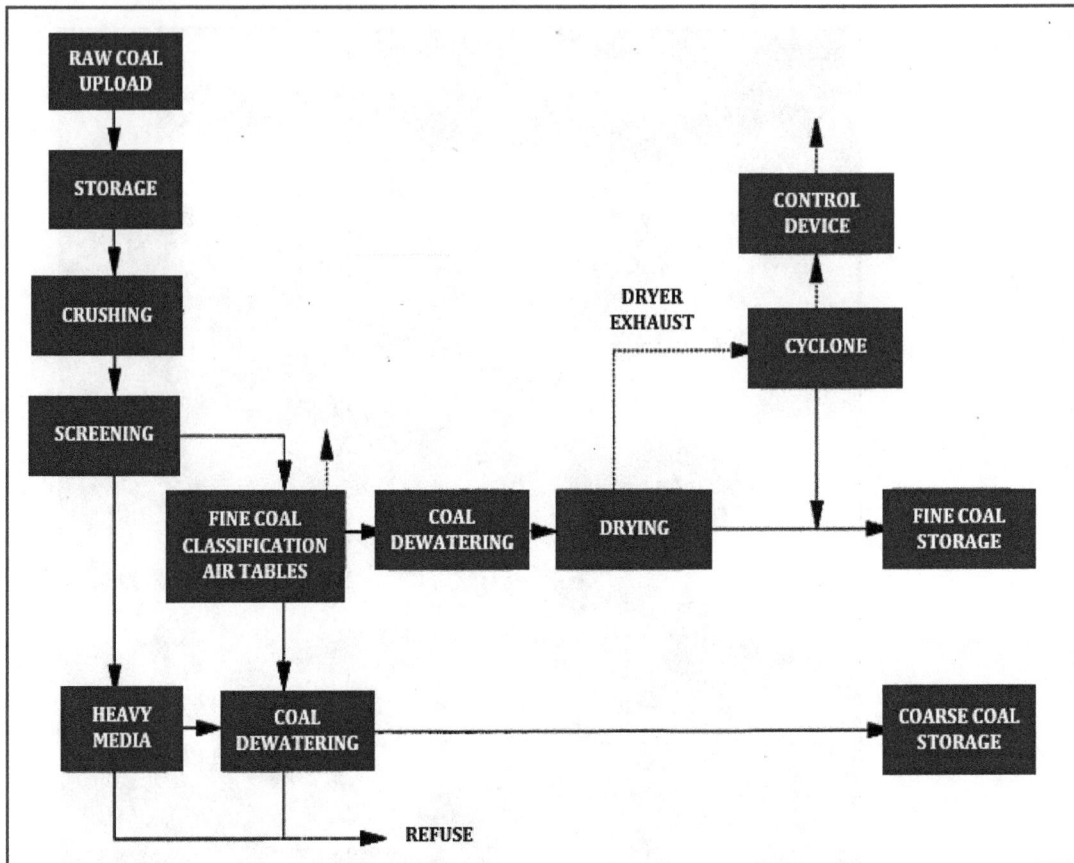

Figure 1.23 Typical coal cleaning plant process flow diagram.

REFERENCES

Afonja A. A. (2017). *Basic Coal Science and Technology*. SineliBooks.

Averitt, P. (1975). Coal resources of the United States. U.S. Geological Survey
   Bulletin 1412, 131p.

Berkowitz, N. (1994). *An introduction to coal technology.* 2nd Edition. Academic Press

BP (2014). *Statistical Review of World Energy.* www.bp.com.

IEA (2013a). *Resources to Reserves, 2013. International Energy Agency.* www.iea.org.

IEA (2013b). *Key Energy Statistics. International Energy Agency.* www.iea.org.

IEA (2014). *Key Energy Statistics. International Energy Agency.* www.iea.org.

IEA (2015). *Key Energy Statistics. International Energy Agency.* www.iea.org.

IEA (2016). *Key Energy Statistics. International Energy Agency.* www.iea.org.

Speight, J. G. (2012). *The Chemistry and Technology of Coal*. Third Edition. CRC Press.

Thomas, L. P. (2013).). *Coal Geology,* 2nd Edition. John Wiley & Sons.

UNECE (1988). "International codification system for medium and high rank coals." United
   Nations Economic Commission for Europe, unece.org.

van Krevelen, D. W (1963). *Coal*. New York, Elsevier.

WEC (2013). *World Energy Resources: 2013 Survey.* World Energy Council.
   www.worldenergy.org.

2   Coal resources of Nigeria

2.1   INTRODUCTION

Coal deposits were found in Enugu area of Eastern Nigeria at the beginning of the 20th century and initial exploitation was for export to fuel British naval ships during the first world war. A railroad was built to link the mines with the port at Port Harcourt to facilitate export and this marked the beginning of the Nigerian railway system which was powered by coal-fired steam locomotives. Subsequently, coal was used for electric power generation, initially to serve the residences of the British expatriates but use was eventually extended to other parts of the country when two power stations were established at Oji river near Enugu and Ijora in Lagos.

2.2   HISTORY OF NIGERIAN COALS

Coal was first discovered in Nigeria in 1909 by a team of British geologists from the Geological survey of Southern Nigeria under the direction of the Imperial Institute of England. A comprehensive survey of the mineral resources of the protectorate began in 1903 and lignite was one of the first minerals found in Ogwashi-Ukwu, Eastern Nigeria and some other areas of the region. With the discovery of sub-bituminous coal in Udi in 1909, interest in lignite was lost and mobilization for developing the Udi mine began. Between 1909 and 1913 more coal outcrops were discovered along the Enugu escarpment through to Onitsha but exploitation efforts focused on the Udi Hill area. Fast-tracked by the 1914-1918 first world war, the mine was in operation by 1915 and the first coal-powered railway was operating between the Udi coal centre and the newly constructed port in Port Harcourt by 1916 (Simpson, 1954).

Further geological work came to a virtual stop during the 1914-18 war but resumed with the establishment of Geological Survey Department in 1922. In the next two decades or so many more outcrops were discovered and evaluated. Initially, the coal industry was a unit of the Marine Department and coal produced was mainly for fueling the British naval ships. Coal was also used by the Nigerian Railways' steam locomotive engines, for steam raising to generate electricity throughout the country, and was also exported to some neighbouring countries of West Africa. Within a short time the industry had established itself as one of the largest employers of labour in the country. The industry was excised from the Marine Department and merged with the Nigerian Railways in 1937 and in 1950, the Nigerian Coal Corporation was established.

The increasing internal demand for coal and bright potential for export to other African countries provided the impetus for more in-depth exploration which resulted in the discovery of more deposits and thicker seams. Production increased steadily, serving the upgraded power plants, the extended railway system, and many new industries including Nkalagu cement plant. Coal export became one of the major foreign exchange earners and, throughout the 1950s, supplied over 70% to the commercial energy requirements of the country.

The decline in coal demand began in 1960 following the discovery of oil in commercial quantities at Oloibiri and the commissioning of the hydroelectric power station at Kainji. Several other reasons for the decline have also been identified and include the absence of a coherent national energy policy which would have promoted the development of a sustainable primary energy mix, lack of clear and specific production-demand targets set for coal in Nigeria, inadequate technological capability to mechanize coal mining operations, and lack of venture capital to invest in coal production. Other problems which confronted the coal industry included poor infrastructural facilities for coal production such as mining, coal preparation, storage, transportation, etc., and dissatisfaction of the miners with their conditions of service (Akarakiri et al., 1991). Henceforth demand declined and the industry finally collapsed in 1999.

2.3   CURRENT STATUS OF THE NIGERIAN COAL INDUSTRY

The output of the Nigeria coal industry peaked in 1959 when nearly a million tonnes of coal was produced. The development of the oil resources marked the beginning of the decline of the coal industry and production had virtually stopped by 1999. Efforts by government between 2001 and 2007 to revitalize the industry through public-private partnerships did not succeed. Of the several partnership agreements between the Nigerian Coal Corporation and private investors, only the one with the Danish-British company, Nordic Industries Limited succeeded in producing about 2000 tonnes of coal from the Okaba opencast mine in 2001. The Nigerian Coal Corporation was dissolved in 2003 to pave way for the privatization of the industry.

Further efforts to privatize the Nigerian Coal Corporation were stalled for several years by a litigation filed by Enugu State Government on ownership of the Nigerian Coal Corporation assets. In 2007, the Mining and Minerals Law was enacted and it empowered the Bureau of Public Enterprises (BPE) of the National Council on Privatization (NCP) to privatize the coal industry, along with many other public enterprises. The coal resources of Nigeria were divided into ten blocks as shown in Table 2.1. Nine of the coal blocks offered for privatization were won by four companies but only two met the deadline for acquiring three of the blocks (Okaba, Ogboyoga I & II,) located in Kogi State. The Federal Government is partnering with HTG/Pacific Energy, a Nigerian-Chinese company and Simang of South Africa to build captive coal-fired power plants at Nsukka and Enugu respectively. The remaining blocks are on offer for privatization.

2.4   COAL RESOURCES OF NIGERIA

Nigeria has abundant resources of mainly soft, low rank sub-bituminous and lignite coals estimated at about 966 million tonnes of proven reserves and nearly twice this value in inferred reserves. Recent exploration work has located several new deposits and there are many outcrops in different parts of the country. Large parts of the country remain unexplored and there are indications that the coal resources may be considerably more than present estimated resources and reserves. The proven deposits are located mostly in the South-Eastern part of the country (Figure 2.1). The details about the reserves are given in Table 2.2. Most of the coal deposits are currently unexploited because of the low demand. The five sub-bituminous coal deposits extend from Enugu to about 160km north to Okaba. The coals are all of the same type but the deposits are not interconnected. The seam thicknesses vary widely, between less than 1 metre and 1.5 metres. Various geological surveys over the last six decades or so have located many more occurrences, some seams up to 2 meters thick, but extremely variable in lateral seam thickness, continuity and quality.

Surveys carried out further north of Enugu in the 1970s located the Obi/Lafia deposit which has significantly different properties compared with all coals found previously. The coal is more mature than the coals in the Enugu-Okaba belt, in contradiction of the earlier conclusion that maturity of the coals decreased from south of Enugu towards the north. The earlier theory of south-to-north maturity profile of Nigerian coals was based on the fact that carbon content of the deposits which is a major indicator of rank and maturity, decreased in the same direction. In fact, the Obi/Lafia deposit has the highest carbon content of all coals found so far in Nigeria.

The estimates of coal and lignite reserves have been based on boreholes drilled in the 1950s and 1960s. Actual resources are believed to be significantly more because no major effort has been made since then to investigate many other deposits that have been located subsequently, with the exception of the Obi/Lafia deposit which was evaluated in the 1970s because of its potential as coking coal for the iron and steel industry. The deposits which have been mined at one time or another are shown in Table 2.3.

Table 2.1 The Nigerian coal blocks. *(Bureau of Public Enterprises, Nigeria).*

COAL BLOCK	STATE	COAL RANK	CONCESSIONAIRE
Okpara	Enugu	Sub-bituminous	
Onyeama	Enugu	"	
Okaba	Kogi	"	Zuma Energy Nigeria Ltd
Orukpa	Benue	"	
Ogboyoga I	Kogi	"	Astra/Shebuel International Ltd
Ogboyoga II	Kogi	"	Astra/Shebuel International Ltd
Ezinmo	Enugu	"	HTG/Pacific Energy Co., Ltd
Inyi	Enugu	"	
Ogwashi-Azagba	Delta	Lignite	
Amansiodo	Enugu	Sub-bituminous	

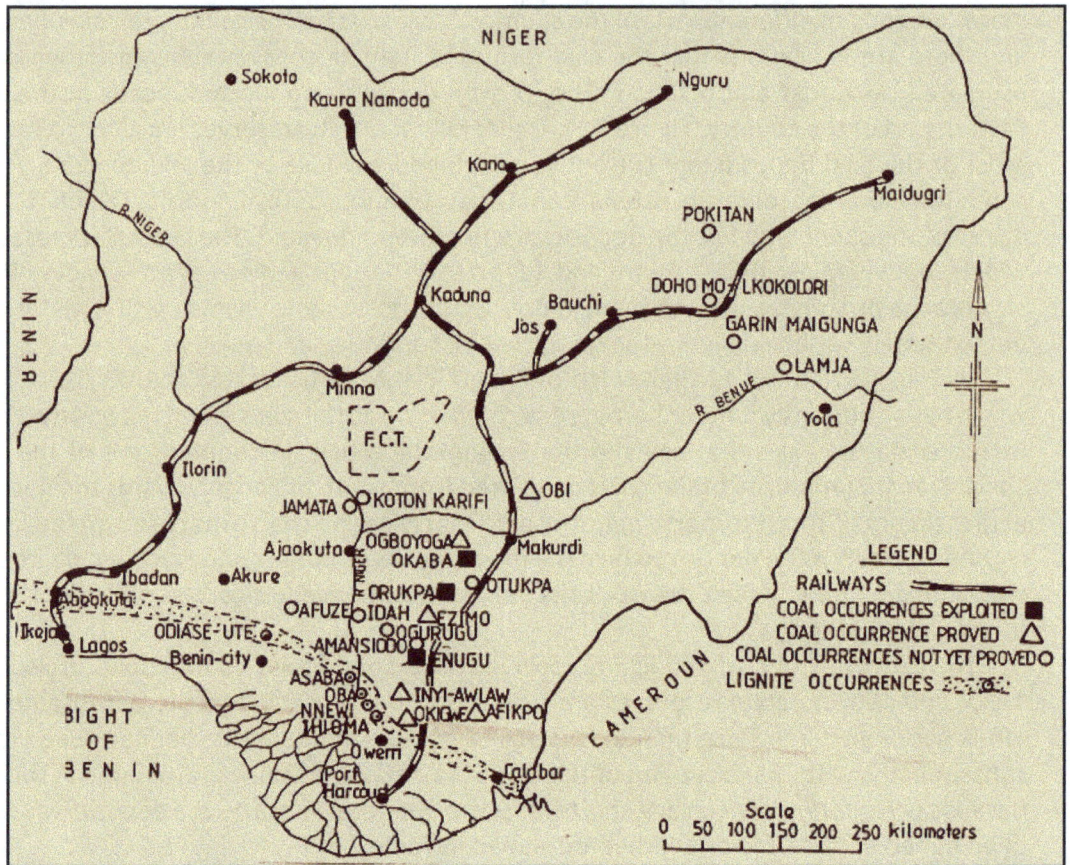

Figure 2.1 Map showing coal deposits of Nigeria. (*Source: Nigerian Coal Corporation*).

Table 2.2 Coal and lignite reserves of Nigeria. (*Compiled from various sources*).

LOCATION	TYPE	RESERVE (million tonnes)
Enugu (Udi)	Sub-bituminous	120.2
Ezimo (Nsukka)	"	17.3
Inyi	"	10.3
Ogboyoga	"	310.5
Okaba	"	190.3
Orukpa	"	70.1
Lafia	Semi-bituminous/Bituminous	32.0
Asaba/Ogwashi	Lignite	64.0
Nnewi-Oba	"	142.0
Obomkpa	"	10.0
Total		966.4

Table 2.3. Coal and lignite reserves of Nigeria. *(Source: M2M Workshop, 2005).*

Mine	Coal Type	Estimated Reserves (mtonnes)	Proven Reserves (mtonnes)	Depth of Coal	Mining Method(s)
Okpara	Sub-bituminous	100	24	180	Underground
Onyeama	Sub-bituminous	150	40	20-80	Underground
Ihioma	Lignite	40	N/A	20-80	Surface
Ogboyoga	Sub-bituminous	427	107	20-100	Surface/Underground
Ogwashi/Asaba	Lignite	250	63	15-100	Surface/Underground
Ezimo	Sub-bituminous	156	56	30-45	Surface/Underground
Inyi	Sub-bituminous	50	20	25-78	Surface/Underground
Lafia/Obi	Bituminous	156	21.42	80	Undergrond
Obi/Nnewi	Lignite	30	N/A	18-38	Underground
Afikpo/Okigwe	Sub-bituminous	50	N/A	20-100	Underground
Amasiodo	Bituminous	1,000	N/A	563	Underground
Okaba	Sub-bituminous	250	73	20-100	Surface/Underground
Owukpa	Sub-bituminous	75	57	20-100	Surface/Underground
Ogugu/Agwu	Sub-bituminous	N/A	N/A	N/A	Underground
Afuji	Sub-btuminous	N/A	N/A	N/A	Underground
Ute	Sub-bituminous	N/A	N/A	N/A	Underground
Duho	Sub-bituminous	N/A	N/A	N/A	Underground
Kurumu	Sub-bituminous	N/A	N/A	N/A	Underground
Lamja	Sub-bituminous	N/A	N/A	N/A	Underground
Garin Maigunga	Sub-bituminous	N/A	N/A	N/A	Underground
Gindi Akwati	Sub-bituminous	N/A	N/A	N/A	Underground
Janaja Koji	Sub-bituminous	N/A	N/A	N/A	Underground

2.5   GEOLOGY OF NIGERIAN COAL DEPOSITS

Central and West African regions have a series of basins believed to have been formed by rifting of the central West African basement complex, beginning at the start of the Cretaceous Period. One of the rifts extends from the northern boundary of the Niger-Delta in southern Nigeria northeastwards for about 1000 kilometres and ranges in width from 80 to 250 kilometres, terminating underneath Lake Chad. The section of this rift located in Nigeria is known as the Benue Trough and terminates at the southern boundary of the Chad Basin. Regionally, the Benue Trough is part of an Early Cretaceous rift complex known as the West and Central African Rift System. In the Upper Cretaceous, the Trough probably formed the link between the Gulf of Guinea and the Tethys Ocean (predecessor of the Mediterranean Sea) via the Chad Basin (Wright, 1985).

Initially, the Trough accumulated sediments deposited by rivers and lakes, but subsided rapidly and was covered by sea during the Late Early to Middle Cretaceous. Towards the end of this period the basin rose above sea level and silted up. Extensive coal forming swamps developed in the low-lying marshes, particularly in the Anambra Basin (Figure 2.2) and eventually, the deep layer of vegetation became buried under coarse sands. At the upper end, the Trough bifurcates into an E-W trending Yola arm and N-S Gongola arm. This late Jurassic-Early Cretaceous to Megastructure is filled with continental and marine sediments, up to 6,500 metres thick in some areas. Literature on the geology and stratigraphy of the Benue Trough is extensive and the mineral resources of the Trough have been the focus of many studies (de-Swardt and Casey, 1963; Carter et al., 1963; Reyment, 1965; Petters, 1982; Obaje et al., 1994; Obaje, 2009; Fatoye et al., 2013b). The Benue Trough is divided into three sub-Troughs: the Lower Benue, Middle Benue and Upper Benue Troughs (Figure 2.3). This division is arbitrary and no clear boundaries have been defined although there are significant differences in the stratigraphy and sedimentology of the sub-Troughs.

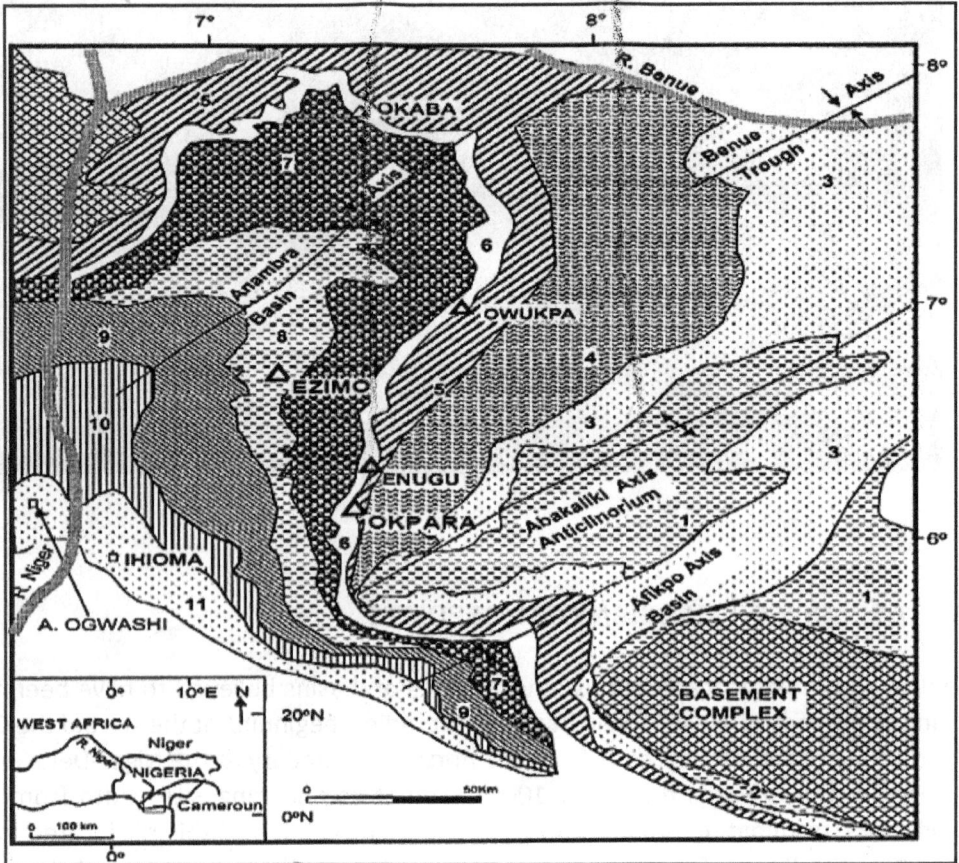

Figure 2.2 Geological map of South-East Nigeria. *(Source: Uzoegbu et al., 2013).*

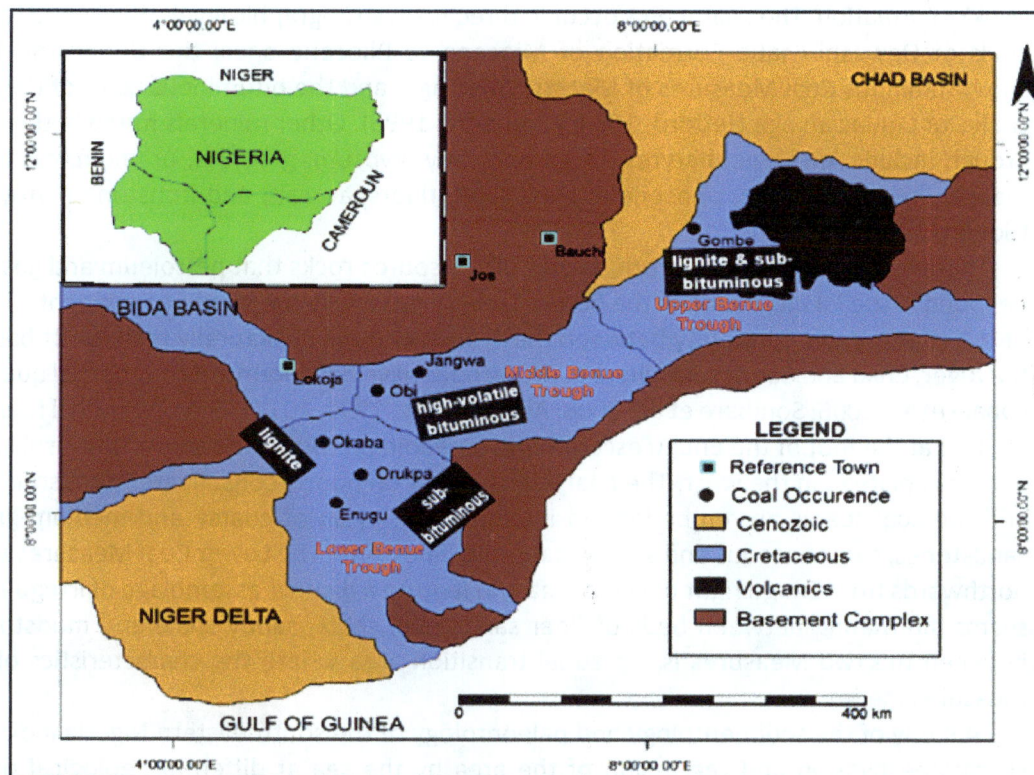

Figure 2.3 Map of the Benue Trough. (*Modified from Obaje et al. (1999); Fatoye et al., 2013a*).

Important differences in geological timelines have been found even within the sub-Troughs. For example, in the Lower Benue Trough, the geology of the Anambra Basin which extends northwards from Enugu through Awka and Okigwe is significantly different compared with that of the adjacent Nkalagu-Abakaliki Basin (Akande and Erdtmann, 1998). The Anambra Basin is believed to have suffered oceanic transgression and subsequent regression during the Campanian-Maastrichtian period as a result of which organic and oceanic sediments were deposited. These oceanic incursions have led to the formation of an extensive coastal area with lagoons and swamps on which a thick succession of Upper Coal Measures and sandstone were deposited. The original sedimentation of the area is believed to have been entirely non-marine.

Geological and stratigraphic evidences link the Anambra Basin, Benue Trough and the Chad Basin, and most of the minerals found so far in Nigeria have been in these areas. The Benue Trough has undergone many phases of cretaceous volcanic activities with peculiar characteristics in its geological setting of excessive tectonic shattering and deformation resulting in tensile and compressive stresses (Famuboni, 1996). A wide range of mineral deposits have been found in the Benue Trough. Most of the coals found so far are located in the Trough, although many outcrops have also been found in other areas outside the Trough. The main coal resources are located in the Cretaceous Anambra and Makurdi Basins, and

Afikpo Syncline. The deposits occur in two levels: the lower Mamu Formation and the upper Nsukka Formation. The coal seams occur in three main stratigraphic levels: the lignite (brown) coals of Ogwashi-Asaba Formation of Miocene to Pliocene ages; the Upper and Lower sub-bituminous Coal Measures of Maastrichtian age; and the bituminous coals of the Agwu shales of Coniacian age (Jefford, 1963; Famuboni, 1996). Other minerals found in the Benue Trough include lead-zinc, barytes, limestone, clay, gypsum, graphite, uranium, cassiterite, manganese, mica, silver, phosphate, glass sand, fluorspar, salt, and ironstone (Fatoye and Gideon, 2013).

There are indications also from studies of the source rocks that petroleum and gas could be found in the inland Basins of the Benue Trough, especially the southern segments, in view of the stratigraphic continuity between the Trough and the structurally related rift basins of the Niger, Chad and Cameroon where hydrocarbons have been found in commercial quantities (Obaje et al., 2006; Sonibare et al., 2008; Akande et al., 1992, 2012). The Upper Coal Measures abound at the foot of the Enugu escarpment, spreading to Udi in the south, and Nsukka area and Igala plateau in the west. The coal deposits rest on poorly consolidated sandstone beds and the coal seams are embedded in intersparsing layers of coarse and medium-grained sandstones, carbonaceous and sandy shale, and limestone. The Lower Coal Measures spread northwards from Enugu right to Ogboyoga and feature a distinct assemblage of irregular coal seams alternating between beds of finer sandstone, shale, sandy shale and mudstone. In between the two Measures is a gradual transition area where the characteristics of both Measures merge.

 A study of the sedimentology and paleontology of the south-eastern Nigeria indicates a successive invasion and regression of the area by the sea at different geological periods ranging from Cenomanian to Miocene. Each invasion deposited a thick series of marine shale, sandstone and limestone. As a result of the regression of the sea, deposition slowed down and eventually ceased. The process of ocean incursion was repeated over and over in different geological time frames and the repeated deposition of marine sediment followed by dry periods probably accounts for the extensive variations in the sedimentation pattern, from continuous to unconformable underlying beds. In later periods the sea gradually became shallow and extensive low-lying coastal lagoons and swampy areas were formed on which successive deposits of thick coal Measures and freshwater sandstones occurred.

It is believed that deposition during this period was predominantly non-marine until the Lower Coal Measures had been deposited but there is evidence that there was some gradual marine encroachment during the deposition of the Upper Coal Measures, probably during the Paleocene and Lower Eocene as indicated by marine intercalations in some of the deposits. By Miocene times extensive coastal areas had become re-established and conditions had become favourable for the deposition of freshwater sediment, carbonaceous beds and lignites (Jefford, 1963). The process of sedimentation of the coal area probably explains the wide variations in the physico-chemical characteristics of the various coals in spite of the closeness of the deposits. For example, Enugu coal is hard and relatively unreactive while Okaba coal is soft, friable and highly reactive.

2.6   COAL DEPOSITS OF THE BENUE TROUGH

The Nigerian coal deposits cover an extensive area of the eastern and northern regions of the country, bound in the north and west by the Benue and Niger rivers respectively, and spreading eastwards to the Enugu escarpment. Many coal outcrops have been discovered in other areas but only a few have been investigated and proven. Lignite and older coal deposits of the sub-bituminous type have been found in the Mamu Formation of the Lower and Upper Benue Troughs. The oldest of the deposits found so far is located in the Obi-Lafia region of the Middle Benue Trough.

The lignite and sub-bituminous coals located within the Mamu Formation of the Lower Benue Trough and in the Gombe Basin are Middle-Campanian to Late Maastrichtian and are believed to have been deposited around the same period (Fatoye and Gideon, 2013). Most lignite deposits worldwide are believed to be Miocene but it is also possible that two plant debris collections that were deposited around the same geological timeline could mature at different rates depending on the nature of the plant debris, the environmental and geothermal conditions of the location. High volatile bituminous coals considered to be the oldest of the coal deposits have been found in the Agwu formation of the Middle Middle Benue Trough, dated Middle Turonian – Early Santonian.

All the coal deposits found in the Lower and Middle Benue Trough are believed to belong to the Mamu Formation which extends northwards from Onitsha in Anambra state through Imo, Kogi, Enugu states up to Obi-Lafia in Nasarawa state. Initially it was thought that all the deposits of the Trough occurred in one formation but an important study by Tattam (1943) identified two clear and distinct coal-bearing formations: the Lower and Upper Coal Measures, separated by a thick series of sandstones. Extensive mapping of both areas has been done (Simpson, 1954; Jefford, 1963) and the Measures fully characterized.

2.6.1 Lignite coals

The youngest coals (lignite) deposits are found in a belt starting from Ogun state in the South-West and extending through Edo, Delta, Anambra, Imo states to Cross River state in the South-South. Although little work has been done on these deposits the first of which was discovered around 1903, they are believed to be of Miocene age and the seams vary considerably in thickness, up to two metres in some areas. The lignite is embedded in coarse sandstone, carbonaceous mudstone, plastic clay, and sand. These deposits have never been exploited commercially.

Coal has also been found in the Gombe Formation, hosted by the Gongola Basin of the Upper Benue Trough. The deposits are believed to have been formed towards the end of the Maastrichtian and are younger than the coals of the Anambra Basin. Some of the seams are as shallow as 15-40 metres and seam thicknesses are sporadic, thin in some locations and as thick as 2 metres in others. The Gombe deposits have not been evaluated hence neither the extent nor the technological quality is known although it is being exploited at the Maiganga mine and utilized by Ashaka Cement Factory located in the area. A coal-fired electric power plant has also been approved for the area recently. The deposit has been variously reported as lignite/sub-bituminous but the results of a recent study on some samples show clearly that it is lignitic (Onoduku, 2014).

2.6.2 Sub-bituminous coals

Extensive coal deposits of sub-bituminous type have been discovered around Enugu, extending northwards for about 160 kilometres to Ogboyoga in the Mamu Formation. The geological successions of all the deposits are similar, although there are significant differences in the physico-chemical properties.

2.6.2.1   *Enugu coal deposit*

Enugu coal occurs in the Mamu Formation (Lower Coal Measures) which comprises alternating sandstones, shales, sandy shales and mudstones, with coal seams and carbonaceous shale at various horizons. The sedimentation shows a rhythmic pattern, with irregular concretions occurring abundantly in the shale and sandy shale (Famuboni, 1996). The Ajali Formation which comprises mainly of poorly-sorted and cross-bedded sand stone overlies the Mamu formation and transits to the higher Nsukka Formation (Upper Coal Measures). Five coal seams of varying characteristics have been identified in the Enugu area. Seam thickness varies across beds from less than ten centimetres to nearly two metres. The No. 3 seam is considered the most economically viable of all and the first mine which opened in Udi in 1915 targeted this seam which is about 200 metres deep. Consequently the deep shaft pillar and stall method of mining was adopted. With increasing demand for coal, the Udi mine could no longer cope and other new mines (Ogbete, Iva valley, Obwetti, Hayes, Prison Creek, Forest Hill, Palm Valley, Ekulu, Okpara, Onyeama, Ribadu, etc.) were opened at different times to work the same seam. Coal from this deposit provided most of the primary energy requirements of the country in the first fifty years of the twentieth century.

2.6.2.2   *Ezimo and Orukpa coal deposits*

Although coal was discovered in two other locations around 1913, the Enugu deposit remained dominant until the late 1940s when the Ezimo and Orukpa deposits about 65-70km north of Enugu at the foot of the Enugu escarpment attracted further interest. Both deposits occupy a higher position in the Mamu Formation (Lower Coal Measures). The geological succession of the area is similar to that of Enugu, with the Mamu Formation in the lower slopes of the escarpment while the upper slopes feature the Ajali Formation. There are many exposures of thin coal seams along the outcrop of the Mamu Formation but only the Ezimo and Orukpa deposits are considered economically viable. The Ezimo deposit is estimated to be about 46 million tonnes. Core samples indicated that the seam thicknesses ranged from 1.5 to nearly 3 metres and the main seam outcrops for about 6.5km along the escarpment. Orukpa is a few kilometers northwest of Ezimo and the coal seam outcrops west of the town, at the base of Enugu escarpment. The deposit is estimated to be about 57 million tonnes. Both deposits can be exploited by opencast or underground mining. The deposits have been privatized and Intensive geological survey is in progress to fully quantify and characterize them for power generation.

2.6.2.3   *Okaba and Ogboyoga coal deposits*

A sub-bituminous coal deposit was found in the Otokpa stream in Okaba in 1930. The deposit is shallow and localized with seam thickness up to 3 metres in some areas. The total reserve is estimated at about 75 million tonnes. In view of the relatively light overburden the deposit has been mined by the open cast technology.

Another deposit was discovered around the same time in Ogboyoga, a few kilometres further north of Okaba. The thickness of the main seam varies from 0.75 to 2.5 metres. The deposit is estimated at about 107 million tonnes and the characteristics are similar to those of Okaba coal but the deposit has never been exploited. Like the Okaba deposit, the seam is relatively shallow and can be exploited by opencast mining. Both Okaba and Ogboyoga have been privatized and are being developed along with a captive power generating plant.

2.6.2.4 *Inyi, Afikpo, Okigwe coal deposits*

Several outcrops have been discovered south and south west of Enugu around Inyi, Afikpo and Nkporo. The area is underlain by sediments of the Uppermost Cretaceous and Lower Ecocene. All the outcrops are in the Nsukka formation (Upper Coal Measures). Only Inyi deposit has been investigated to some extent. Borehole drilling put the reserve at around 10 million tonnes. The deposit has been privatized and geological investigation of the area has commenced. It is expected that the deposit will supply a coal-fired power plant to be cited at Nsukka.

2.6.3 Bituminous coals

The Mamu Formation marks the regressive phase of the Upper Campano-Maastrichtian transgressive cycle and is completely non-marine (Obaje, 2009). In contrast, formation of Obi-Lafia coal deposits were interrupted by marine oscillatory incursions and this explains the significant differences between these coal deposits and the Mamu Formation coals, in particular, the high sulphur content, a well known characteristic of coal deposits that had been subjected to marine incursions or had been formed from peat accumulated in marine ponds that eventually dried up or regressed. The Obi-Lafia deposit is located about 230km south of Jos in about forty horizons. Detailed exploration of the deposit was carried out by the Steel Raw Materials Exploration Agency (SRMEA) in the late 1970s and early 1980s. The deposit occurs within the Agwu Formation which is predominantly shale, limestone and sandstone. Thirty two seams were identified, with a total estimated deposit of 156 million tonnes but only three seams: 12, 13 and 20 were considered commercially viable.

The Obi-Lafia deposit is sporadic and most of the seams are discontinuous, too thin or of too poor quality to justify detailed investigation. Also, all the seams are deep and the geological structure of the area makes the technical feasibility of underground mining doubtful. Two seams which are relatively thick and could be mined economically if the quality is good have been investigated in depth. Further survey of the area has located more deposits which are currently being investigated. Available data from the analysis of 137 borehole samples from 32 seams (Afonja 1972, 1975, 1979) show that the deposit is of higher rank than any other coal found so far in Nigeria and belongs to the high-volatile bituminous rank. However the seams are sporadic and thicknesses vary widely. Also the coal seams are intersparsed with and intruded by pyrites layers which evidently had been associated with the coal deposit for a very long time, considering the fine dissemination of sulphur in the coal.

The geology of the deposit indicates that mining may not be economical. Furthermore, the seams are interspersed with thick layers of mineral shale and pyrites. The sulphur content is very high and the coal does not respond well to washing. The sulphur content is too high for metallurgical coking but is acceptable for power generation. However, expensive pre or post combustion sulphur removal treatment would be required in order to meet the increasingly stringent environment pollution regulations. Of the 36 seams investigated only seams 12 and 13 are considered mineable and recovery is expected to be low (around 25% of the estimated 20 million tonnes in the two seams) in view of the disturbed geological formation in which the deposit is embedded. Heavy faults both normal and reversed, with throws ranging from 8 metres to 125 metres abound and mining would require the adoption of fully or semi-mechanized technology which would be very expensive and unjustifiable for such a small recoverable deposit. However, the deposit may be potentially suitable for in-situ underground gasification. Coal deposits have been found in Adamawa, also in the Middle Benue Trough. Although the existence of the deposit has been known for some time, neither the extent of the deposit nor the technical quality has been investigated in any depth. However, available geological data and preliminary analysis on borehole samples indicate that the deposit has similar characteristics to the Obi-Lafia deposit located in the same Trough.

2.6.4 Other coal resources of Nigeria

Coal deposits have also been found in as many as thirteen states of Nigeria. Coal seams about 2 metres thick were discovered near Doho in Gombe and seams as thick as 2-4 metres were also discovered in Garin Maigangu, south of Gombe. There is also geological evidence that large coal reserves may be present in greater depths than so far investigated in the Mamu Formation around Amansiodo in Enugu State. Other deposits have been identified in Koton-Karfi, Kogi State and Ute in Ondo State. The geological and stratigraphic history of the Anambra and Benue Basins is potentially favourable to mineral formation. However, only small areas of both Basins have been mapped or subjected to any systematic geotechnical investigation, hence the optimism that many more deposits of coal and other minerals will be discovered in the areas.

2.7 RANKING OF NIGERIAN COALS

Many coal classification systems have been developed and the most commonly used have been discussed in chapter 1. The two most popular are the British National Coal Board (UK-NCB) and American Society for Testing and Materials (ASTM) systems which are in use in many coal-producing countries in different modified forms. The British system uses volatile matter and caking value for classification while the ASTM system uses fixed carbon, calorific value and caking properties. The Suggate ranking system is one of the latest and was developed specifically for New Zealand coals but has been found to be applicable to many other coals. The Suggate system uses volatile matter and calorific value for ranking coals.

All Nigerian coals discussed above have been placed in one of three ranks (bituminous, sub-bituminous or lignite), using the British system but would fall in different rank groups depending on the classification system adopted. The ranks of the major coal deposits under three different systems are shown in Table 2.4. Data on Nigerian coals from the Obafemi Awolowo University Coal Research Laboratory (CRL-OAU) database have been plotted on the Suggate graph as shown in Figure 2.4, along with twenty foreign coking coals (low, medium and high volatile coking coals) (Afonja, 1992). All the coals in the Anambra trough fall in the sub-bituminous rank while Ogwashi and Obi-Lafia coals fall in the lignite and high-volatile bituminous rank respectively.

The technological properties of some of the Nigerian coal deposits have been investigated in some depth but many remain largely untested. Most of the coals are suitable for electric power generation but not for metallurgical applications. Also the potential for export is low because of the low heat value per unit weight and the propensity for oxidation and self-ignition in storage. However export to some neighbouring countries may be economically and technically feasible. Available data is presented in a later chapter.

Table 2.4   Ranking of Nigerian coals under different systems.

Coal deposit	Classification systems		
	UK-NCB	A.S.T.M	Suggate
Enugu	902 (non-caking)	20 Bituminous (non-agglomerating)	11.2 Bituminous
Orukpa	902 (non-caking)	20 Bituminous (non-agglomerating)	9.2 Sub-bituminous
Okaba	902 (non-caking)	20 Bituminous (comonly-agglomerating)	8.5 Sub-bituminous
Lafia	401 (strongly-caking)	20 Bituminous (non-agglomerating)	11.7 Bituminous
Ogwashi	902 (non-caking)	20 Bituminous (non-agglomerating)	5.3 Lignite

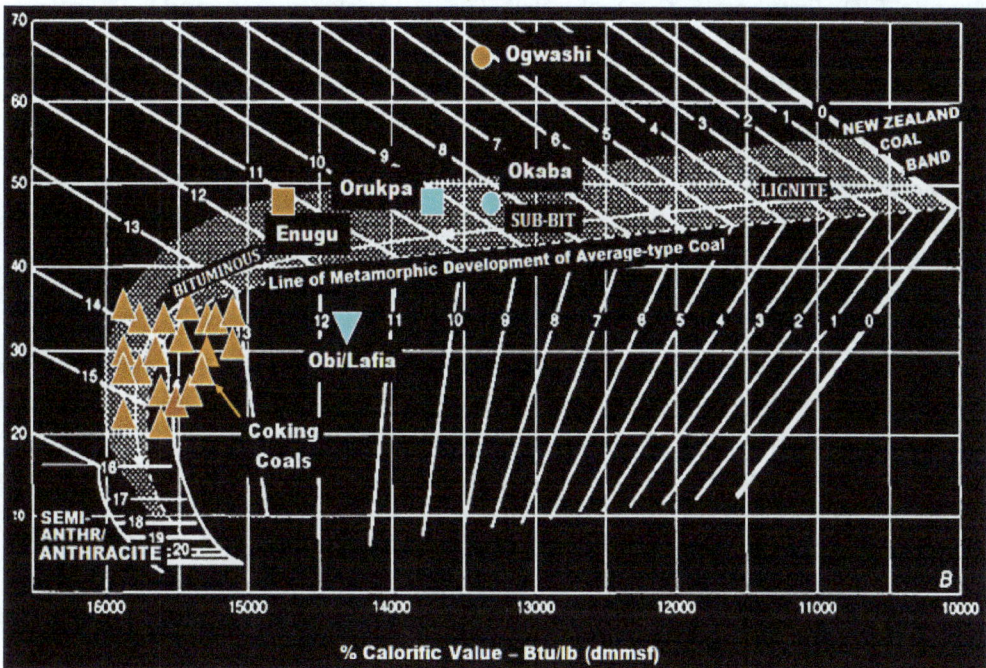

Figure 2.4   Suggate ranking of some Nigerian coals. *(Afonja, 1994).*
[Twenty foreign coking coals from the database of Coal Research Laboratory, OAU, Ile-Ife (CRL-OAU) are shown also].

REFERENCES

Afonja, A. A. (1972). "An assessment of the coking quality of Nigerian coal seams." ENE/B/W/3. Report No. IFE/CHE/CP/1. For the Nigerian Steel Development Authority. (N.S.D.A.).

Afonja, A. A. (1975). "Analytical and coking studies of Obi/Lafia coal deposit. Report No. IFE/CHE/CP/3. For the N.S.D.A.

Afonja, A. A. (1976). "Petrography of Nigerian Coals." Research Report No. IFE/CHE/CP/05, Coal Research Laboratory, Ife.

Afonja, A. A. (1977a). Washability charts for some Nigerian coals. Journal of Mining and Geology, 14(1), 47-49.

Afonja, A. A. (1977b). Feasibility of cleaning Iva Mine coal : sink and float tests. Report No. IFE/CHE/CP/8 . For the Nigerian Coal Corporation.

Afonja, A. A. (1979). Further studies on the Obi/Lafia coal deposit - Phase II. Report No. IFE/CHE/CP/11. For the N.S.D.A.

Afonja, A. A. (1994) Coal: Resources, Development and Utilization. NSCh.E/UNDP Conference, Lagos, Nigeria.

Afonja A. A. (2017). *Basic Coal Science and Technology*. SineliBooks.

Akande, S. O, Hoffknecht, A and B.D. Erdtmann (1992). "Rank and petrographic composition of selected upper cretaceous and tertiary coals of Southern Nigeria." *Int. J. Coal Geol.*, 71: 209-224.

Akande, S. O. and B. D. Erdtmann (1998). "Burial metamorphysm (Thermal Maturation) in Cretaceous sediments of the Southern Benue Trough and Anambra Basin, Nigeria." *AAPG Bulletin*, 82(6), pp. 1191-1206.

Akande, S. O., Egenhoff, S. O., Obaje, N. F., Olo, O., Adekeye, O. A., and B. D. Erdtmann (2012). "Hydrocarbon potential of Cretaceous sediments in the Lower and Middle Benue Trough, Nigeria: Insights from new source rock faces evaluation.". *Journal of African Earth Sciences*, Vol. 64, pp. 34-47.

Akarakiri, J. B., Afonja, A. A. and E. C. Okejiri (1991). "A study of coal production in Nigeria." *Materials and Society* Vol/Issue: 15:4.

Carter, J. D., Tait, E. A. and W. Barber (1963). "The Geology of Parts of Adamawa, Bauchi and Bornu Provinces in North-Eastern Nigeria. Geological Survey of Nigeria.

de-Swardt, A. M. J. and O. P. Casey (1963). *The coal resources of Nigeria*. Geological Survey of Nigeria Bulletin No. 28.

Famuboni, A. D. (1996). "Maximizing exploration of Nigeria's coal reserves." In Okolo and Mkpadi (Eds.) Nigerian Coal: A resource for energy and investments. Raw Materials Research and Development Council, Nigeria.

Fatoye. F. B. and Y. B. Gideon (2013a). "Geology and mineral resources of the Lower Benue Trough, Nigeria." *Advances in Applied Science Research, 4(6), pp. 21-28.*

Fatoye, B. B., and Y. B. Gideon (2013b). "Appraisal of the Economic Geology of Nigerian Coal Resources." Journal of Environment and Earth Science, Vol. 3, No 11, pp. 25-31.

Jefford, G. (1963). "Physical and chemical characteristics of Nigerian coals." Geological Survey of Nigeria Bulletin No. 28.

M2M Workshop – Nigeria (2005). Nigeria's Country Report on Coal Mine Methane Recovery and use. Presented at the Methane to Markets Regional Workshop, Beijing, China.

Obaje, N. G., Ligouis, B., and S. I. Abba (1994). "Petrographic composition and depositional environments of Cretaceous coals and coal measures in the Middle Benue Trough of Nigeria." International Journal of Coal Geology, Vol. 26, pp. 233-260.

Obaje, N. G., Abaa, S. I., Najime, T., and C. E. Suh (1999). *African Geosciences Review*, Vol. 6, Pp. 71-82.

Obaje, N. G. (2009). *Geology and Mineral Resources of Nigeria*. Springer Dordrecht Heidelberg, London, New York. 221 pp.

Onoduku, U. S. (2014). "Chemistry of Maiganga Coal Deposit. Upper Benue Trough, North Eastern Nigeria." *Journal of Geosciences and Geomatics*, 2014 2 (3), pp 80-84.

Petters, S. W. and C. M. Ekweozor (1982). "Petroleum Geology of Benue Trough and Southeastern Chad Basin, Nigeria." Bulletin, American Association of Petroleum Geologists, Vol. 66, pp. 1141-1149.

Reyment, R. A. (1965). *Aspects of the Geology of Nigeria: The Stratigraphy of the Cretaceous and Cenozoic Deposits*. University of Ibadan Press.

Simpson, A. (1954). "The geology of part of Onitsha, Owerri and Benue Province. The Nigerian Coal field." Geology Survey Nigeria Bull. 24.

Sonibare, O., Alimi, H. Jarvie, D. and O. A. Ehinola (2008). "Origin and occurrence of crude oil in the Niger Delta, Nigeria." Journal of Petroleum Science and Engineering, Vol. 61, pp. 99-107.

Tattam, C. M. (1943). "A Review of Nigerian Stratigraphy. Research and Educational Development of the Geological Survey of Nigeria.", 26-27.

Uzoegbu, U. M., Uchebo, U. A. and I. Okafor (2013). "Lithostratigraphy of the Maastrichtian Nsukka Formation in the Anambra Basin, S. E. Nigeria.". , Vol. 5(5), pp. 96-102.

Wright, J. B., Hastings, D. A., Jones, W. B. and H. R. Williams (1985). *Geology and mineral resources of West Africa*. Allen and Urwin, London.

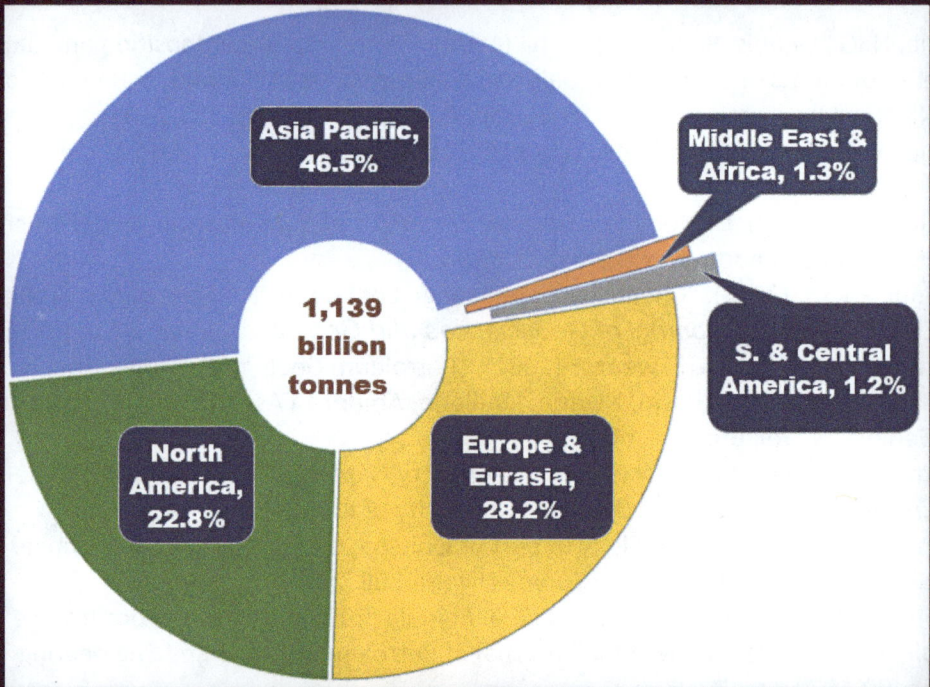

Global coal reserves by region
(bp.com, 2017)

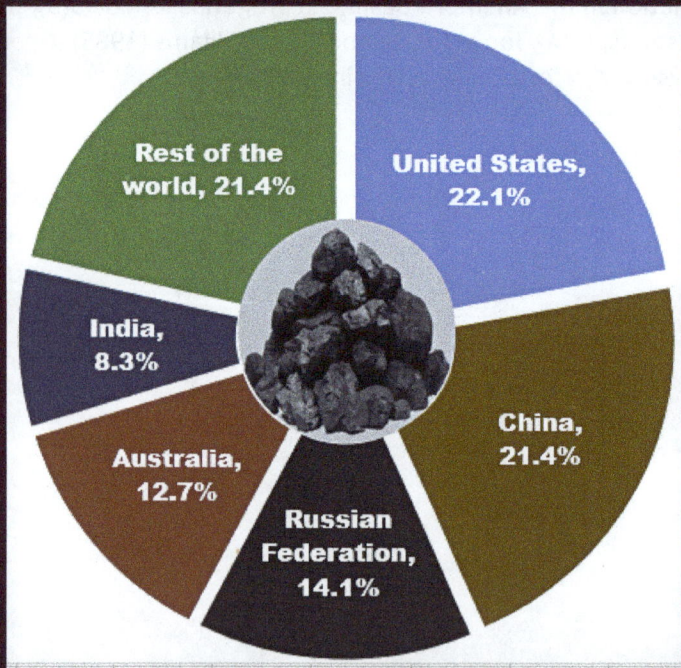

Five countries hold 76% of global coal reserves
(bp.com, 2017)

3

Physico-chemical and petrographic properties of Nigerian coals

3.1 INTRODUCTION

Analytical tests have been carried out on Nigerian coals for nearly a century but there are still no reliable data available on the physical, chemical and thermal properties on which reliable decisions on technological utility can be based. Most of the tests were carried out on borehole samples and outcrops neither of which is truly representative of a coal deposit. The organic and inorganic precursors and geological conditions of burial vary widely hence all important characteristics of the resultant coal may vary significantly and widely from point to point in a coal seam and for seams in the same deposit. In effect, preparation of fairly representative sample is crucial in order to obtain reliable and useful information on the characteristics of any coal deposit. Furthermore, coal weathers on exposure to the atmosphere and significant changes in physical and chemical characteristics may occur within a few weeks of exposure, especially in low-grade coal ranks to which Nigerian coals belong. Also, most of the analysts did not specify the basis of analysis (as received, dry, water-free, dry-ash-free, dry-mineral-matter-free) or the analytical equipment and standards used. Values of a coal property can vary by up to 30% depending on the basis of analysis.

It is important to note therefore that, despite an attempt to integrate and reconcile all available data on Nigerian coals, much of the data presented in this chapter may not be reliable, hence information on the sources of samples is given to indicate the extent of reliability. Virtually all published results on Nigerian coals since the mines were closed in the early 1990s emanated from tests on hand-picked outcrops. Coals picked from the environment of closed mines are heavily weathered and do not represent the coal deposit. The wide variability of available data on the same coal deposits is demonstrated in Table 3.1 which was compiled from data obtained from over twenty sources. Data was obtained from most of the tests carried out on borehole samples, mine cart samples, samples handpicked in closed mines, or outcrops which were probably heavily altered by weathering.

3.2 ORIGIN AND PETROGRAPHY OF NIGERIAN COALS

Coal is formed from the accumulation of plant remains under favourable environmental conditions. These deposits known as peat undergo transformations which involve biochemical, physical and chemical changes through four major stages from peat to mature coal, a process which takes millions of years.

Table 3.1 Chemical composition of Nigerian coals compiled from many sources.

Chemical Parameter	Coal Source				
	Enugu	Ezimo	Orukpa	Okaba	Ogboyoga
Moisture (%)	2.5-12.2	11.1-13.8	10.8-17.2	7.9-12.7	12.6-15.9
Ash (%)	4.1-25.5	5.9-13.1	7.1-12.3	6.8-11.1	8.1-9.5
Vol. Matter (%)	31.6-42.1	38.7-42.1	34.9-38.1	39.8-43.2	37.9-43.1
Fixed Carbon (%)	32.9-47.1	36.7-42,3	38.7-41.7	37.9-42.5	37.6-39.4
Carbon (%)	62.1-69.6	59.9-64.1	57.5-59.7	56.4-61.2	51.8-58.9
Hydrogen (%)	4.1-5.3	1.2-4.9	3.4-3.9	3.5-3.8	3.6-4.9
Nitrogen (%)	1.1-2.3	1.4-1.7	1.1-1.8	1.4-1.8	1.3-1.6
Sulphur (%)	0.4-1.8	0.5-0.9	0.2-0.5	0.5-0.8	0.4-0.8
Oxygen (%)	8.2-14.1	11.0-13.1	10.9-14.4	14.9-17.9	14.8-21.1
Calorific Value (MJ.kg)	22.1-28.7	24.6-25.8	25.1-26.1	22-7-24.6	22.8-24.1

The processes involved in the transformation of plant debris into mature coal and the various stages of the coalification process have been treated in some depth in Chapter 1 (see also Afonja, 2017). The origin and properties of different types of coal depend on a complex interaction of factors including the nature of the original plant debris, the physico-chemical properties of the pond (mire) in which it is buried, the geological history of the site, the topography and physiography of the area, and the prevailing climate.

3.2.1   Origin of Nigerian coals

Available information on the geology of the Nigerian coal deposits has been presented in Chapter 2, but very little work has been done to relate this factor and numerous other relevant factors to the structure, physical, chemical and technological properties of the coals. However, available technical data on the coals show that they fall between the second and fourth stages of transformation. Ogwashi-Uku and Gombe deposits are in the lignite stage while the others are in transition from low-rank sub-bituminous to medium-to-high volatile bituminous coals (stages 3 and 4). The coal deposits in the Enugu-Okaba zone have been classified as belonging to the Lower Coal Measures of Campanian-Maastrichtian age. Many other outcrops have been located west of Enugu which belong to a higher formation, the Upper Coal Measures ranging in age from Uppermost Cretaceous to Paleocene. The lignite deposits in Ogwashi-Uku near Asaba have been classified as of Miocene age (de Swardt, 1963). Work by Estel (1976) on Enugu, Orukpa and Okaba deposits also indicated that the coals are of Tertiary origin.

3.2.2   Petrography of Nigerian coals

The structure of coal as determined by petrography has been discussed in some depth in a companion book (Afonja, 2017, Chapter 5). It is important to note however that the organic and inorganic micro-constituents of coal give vital information on the origin, the nature of the peat-forming plant debris, and the stage of coalification, all of which are useful in characterizing the coal and determining the potential uses. There are three basic groups of macerals, the vitrinite group derived from coalified woody tissue, the liptinite group derived from the resinous and waxy parts of plants and the inertinite group derived from charred and biochemically altered plant cell wall material. The most important sub-macerals in each group are shown in Table 3.2.

Table 3.2 Coal maceral groups and sub-groups.
(ICCP, 1998, 2001).

Maceral Group	Maceral	
Humanite	Telinite	Collotenite
(R₀max ≤ 0.5)	Vitrodetrinite	Collodetrinite
Vitrinite *(R₀max ≥ 0.5)*	Gelinite	Corpogelinite
	Alginite	Bituminite/Amorphinite
Liptinite	Liptodetrinite	Sporinite
	Cutinite	Suberinite
	Resinite	Chlorophyllinite
	Fusinite	Semifusinite
Inertinite	Funginite	Macrinite
	Micrinite	Inertodetrinite

The three organo-petrographic maceral groups are relatively homogeneous and, by their physico-chemical characteristics and relative proportions in coal, largely determine the technical properties and utilization of coal. All coals also contain inorganic materials in varying proportions, classified as mineral matter. The mineral matter group is heterogeneous and may contain a wide range of inorganic minerals, depending on the geological history of the coal. The chemical nature of the mineral matter, morphology and relative proportion to the organic content affect significantly the technological properties and economic value of coal.

3.2.2.1 *Maceral composition of coals*

The classification of the organic macerals is based either on similar origin and/or differences in preservation. There are significant differences in chemical and physical properties such as morphology, elemental composition, moisture content, hardness, density, reactivity. These characteristics also change in the course of diagenesis and coalification (Bustin et al., 1985; Stach et al., 1982; Diesel, 1992). The humanite derives its name from the mode of formation which involves humification of remains of cell walls, woody tissue of stems, branches, leaves, roots of plants and the precipitated gels from these materials. These biochemical processes vary in intensity and progress at different rates in the process of coalification. Remnants of the organic matter are prominent in low-grade coals.

Macerals are defined according to their grayness in reflected light: liptinites are dark gray, vitrinites are medium to light gray, and inertinites are white and can be very bright. The precursor to vitrinite is humanite and transformation to vitrinite progresses over millions of years, with the gradual disappearance of the woody morphology. The vitrinite group is the most abundant in many coals and commonly make up 50 to 90% of the total microscopic contents. However, some coals, for example, western Canadian coals are vitrinite poor and are dominated by inertinite, often making up 50 to 70% of some coals in the region. Matrix vitrinite is almost always the most abundant maceral present and makes up the groundmass in which the various liptinite and inertinite macerals are dispersed. It has a uniform gray color and it is always anisotropic.

The microstructure of humanite often shows preserved intact botanical cell structures visible to various extents and often in isolated cells, or fine humic fragments. It is only present in young coals of the lignite and sub-bituminous ranks. Humanite becomes vitrinite which is the most dominant maceral in mature coals. On pyrolysis, vitrinite may be reactive or non-reactive depending on maturity as well as type of coal. Relatively few coals, all in the bituminous rank contain reactive vitrinite, and they are referred to as *caking* or *coking* coals. The liptinite group derived from the resinous and waxy parts of plants such as spores, cuticles, and resins, which are resistant to weathering and diagenesis. They are composed of hydrogen-rich hydrocarbons derived from spores, pollens, cuticles, and resins in the original plant material. When present in coal, they tend to retain their original plant form. During pyrolysis such as coking, macerals in the liptinite group devolatilize to produce gases and oily tars.

The inertinite macerals derived from plant material that has been strongly altered, degraded and carbonized in the peat stage of coal formation, possibly by ancient fires in coal swamps, volcanic eruptions, moldering, fungal attack, biochemical gelification and oxidation of the same plant tissues which, when not subjected to these conditions, are humified and eventually form huminite/vitrinite. Inertinites are mainly oxidation products of other macerals and are consequently richer in carbon than liptinites or vitrinites. Also, inertinites are predominantly unreactive during pyrolysis, hence are often grouped with mineral matter in coking technology. However, inertinites are rich in carbon and combustible, unlike mineral matter. One of the main distinguishing features between low rank, medium rank and high rank coals is the extent of colonies of liptinites and inertinites embedded in the predominant vitrinite blocks. Vitrinite in low grade coals features, preserved botanical cell structures, cell walls and fragmented plant remains. These features reduce with coalification and may not be present in high rank coals.

3.2.2.2   *Reflectance of coal macerals*

Coal macerals are optically homogeneous organic constituents of coal, possessing distinctive physical and chemical properties. Each of the three groups has subgroups with distinct morphologic features and are different shades of gray. Vitrinite is the dominant maceral in many coals but, in some coals inertinite is the predominant maceral. Liptinite may be completely absent or present in proportions up to about 15%. Coal macerals are identified primarily on the basis of differences in reflectance under incident light. Carbon-rich inertinite has the highest reflectance, hydrogen-rich liptinite the lowest, with oxygen-rich vitrinite in between. The reflectance of vitrinite is a major indicator of coal maturity and numerous investigations have established correlations between the variable and coal rank (Figure 3-1). The macerals also differ significantly in morphology and chemical compositon. However, as coalification proceeds towards full maturity maceral properties converge and become virtually indistinguishable in anthracite. (Figure 3.2).

3.2.2.3   *Petrography of Nigerian coals*

A few studies on the petrography of some of the Nigerian coals have been carried out (Afonja, 1975, 1976, 1979; Estel, 1976; Wuyep et al., 2012; Uzoegbu et al., 2014). The two major parameters of interest in these studies were the maceral composition and the reflectance of the vitrinite content of the coals, both of which are vital for ranking the coals. They also give vital information on the potential areas of use.

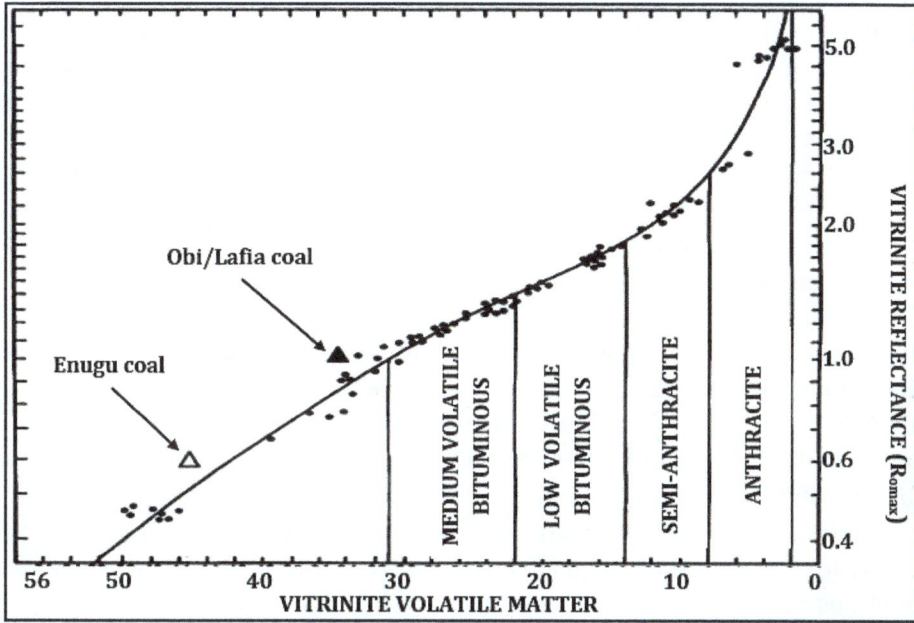

Figure 3.1 Relationship between coal rank and vitrinite mean maximum reflectance (Romax). *(Afonja, 1975, 1979; Bustin et al., 1985).*

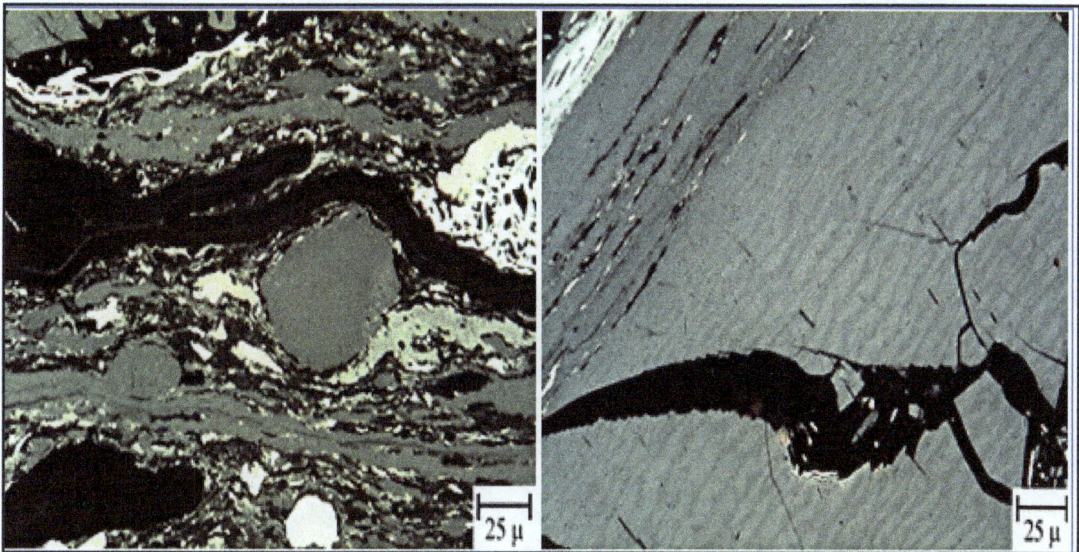

(a) Low-grade coals
(Lignite/Sub-bituminous coals)

(b) Mature coal
(anthracite)

Figure 3.2 Micrographs of (a) low-grade coal showing distinct woody tissue and colonies of liptinite (dark) and inertinite (white) embedded in vitrinite (gray) (b) vitrinite in anthracite with very few colonies of other macerals. *(Crellings, 2017).*

The microstructures of the four major Nigerian coal deposits are presented in Figures 3.3-3.5. All the coals display fungal spores typical of Tertiary origin low rank coals. All the coals contain resinite, most prominent in Enugu coal and are present in dark, round form. Orukpa coal contains sclerotia and cuticles which to some extent reflects the nature of the plant debris from which the coal was formed. Ulminite is prominent in Okaba coal, with shrinkage fissures but the maceral is also present in Enugu and Orukpa coals in smaller quantities. This is unusual since ulminite normally features in lignites. Enugu coal contains significant amounts of clay minerals, intimately intergrown with the coal. Orukpa and Okaba coals also contain clay but in relatively small quantities and in discrete isolated grains. Small quantities of quartz and pyrites are also present in the three coals.

Obi/Lafia coal features highly porous vitrinite and some fusinite in low concentration, high percentage of finely disseminated ash-forming shale, clay, silt, and pyrites intimately embedded in the vitrinite which is the main reactive component of coal. All Nigerian coals studied so far, with the exception of Obi/Lafia coal feature highly structured vitrinite with prominent islands of dark remnants of plant tissue and preserved botanical cell wall structures (liptinite), and highly reflective open cell wall structures (inertinite). These features are characteristic of lignites, sub-bituminous coals and high-volatile bituminous coals, but progressively vitrinize as coalification advances. The features are less prominent in low volatile bituminous coals and may be completely absent in anthracite. The vitrinite in Obi/Lafia coal is more amorphous, indicative of higher maturity compared with the other coals.

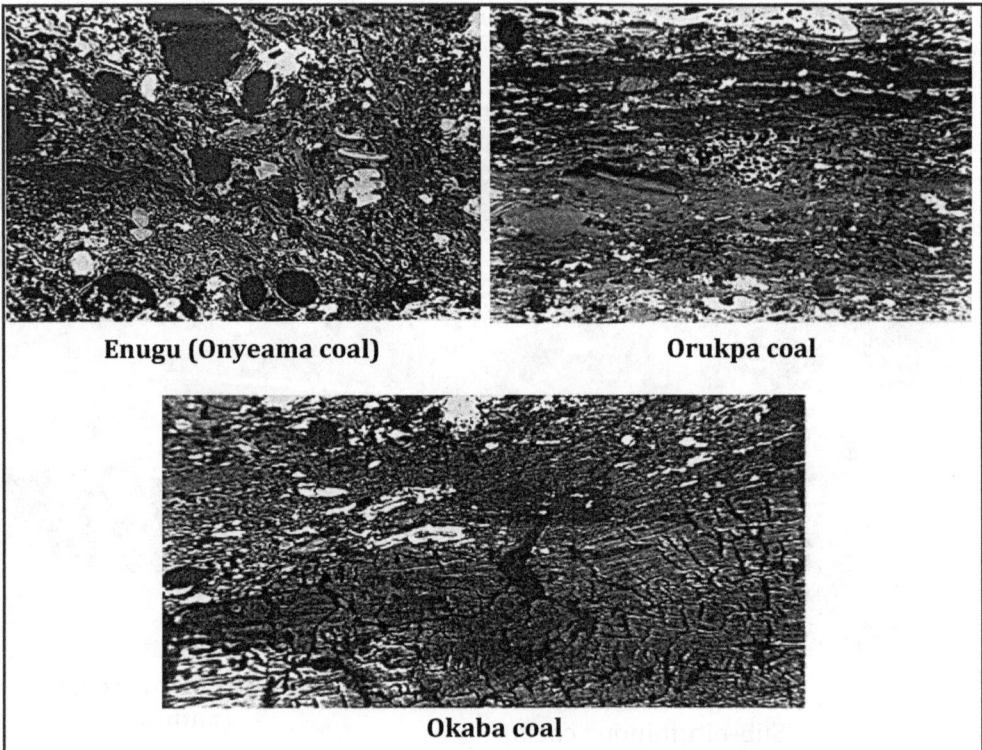

Enugu (Onyeama coal)   Orukpa coal

Okaba coal

Figure 3.3 Micrographs of three coals from the Lower Benue Trough. Enugu coal shows dark, round resinite and white inertinite; the prominent features in Orukpa coal are exinite (dark) and inertinite (white); Okaba coal has similar microconstituents compared with Orukpa coal but extensive fractures are visible, indicating the fryabiity of the coal. *(Source: Roe, 1982).*

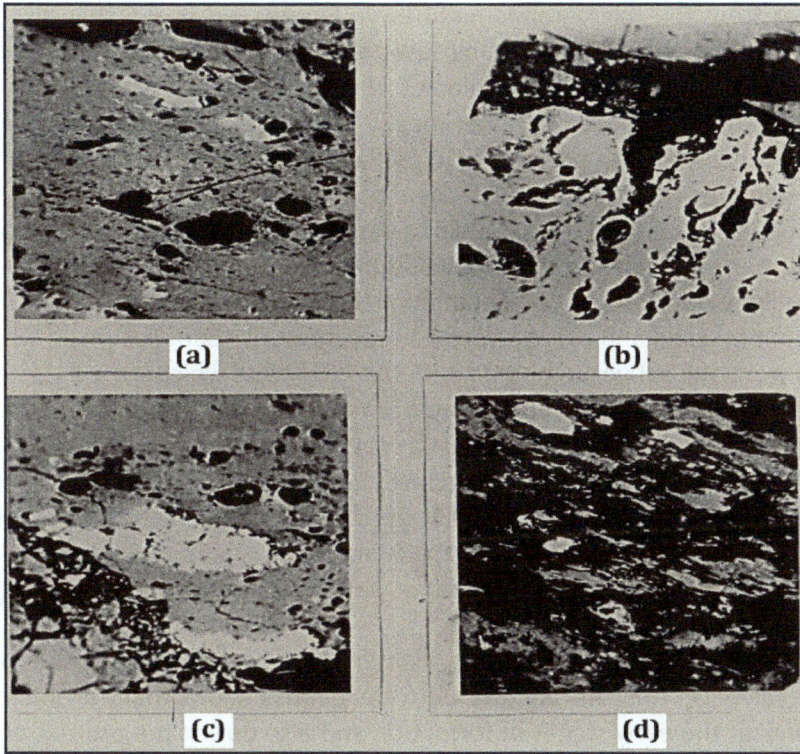

Figure 3.4 Micrographs of Obi/Lafia coal showing: (a) highly porous vitrinite, (b) low-concentration fusinite, (c) pyrites, (d) fine clay and silt. *(Afonja, 1979).*

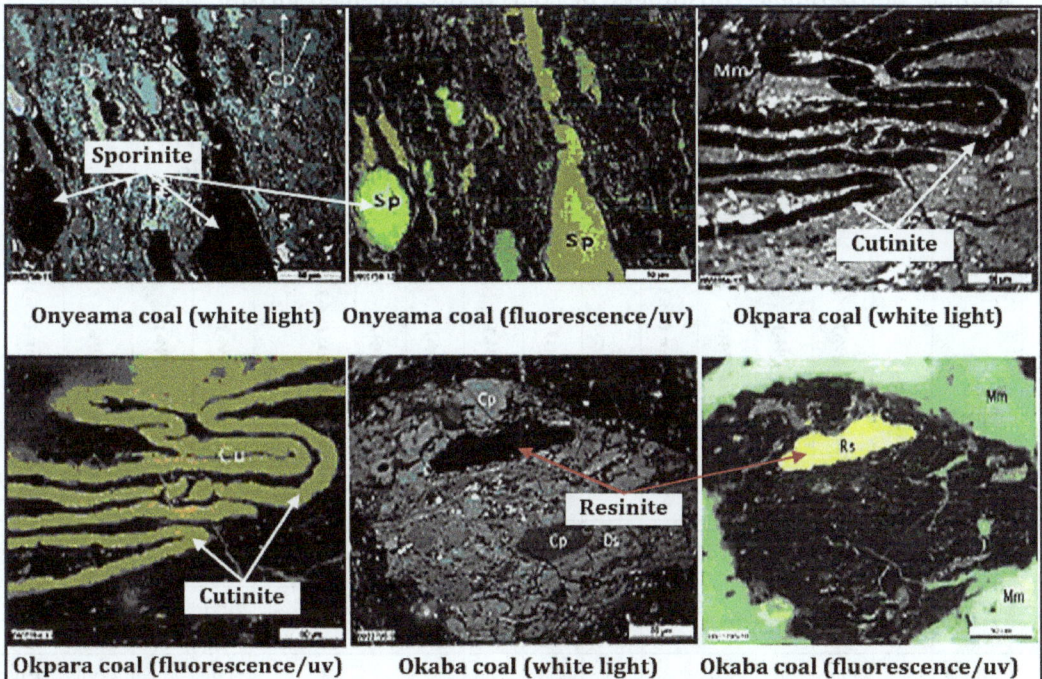

Figure 3.5 Micrographs of Onyeama, Okpara and Okaba coals in white light and fluorescence/uv. *(Adapted from Uzoegbu et al., 2014).*

It is clear from the micrographs featured in Figures 3.3 to 3.5 that all the coals with the exception of Obi/Lafia coal are young, low-grade coals. Humanite is the predominant maceral and the micrographs show large colonies of preserved intact plant cell structures. There are also prominent colonies of liptinite, mostly embedded in massive colonies of inertinite. In contrast, other macerals present in Obi/Lafia coal are embedded in largely coherent though porous vitrinite. This is strong evidence that the deposit is more mature than all the others and humanite has largely transformed to vitrinite. The micrographic features of the coal also place the deposit at the low end of the bituminous rank.

Available data on the petrography of Nigerian coals are presented in Table 3.3. Tests carried out by Estel and Afonja were on representative samples taken while the mines and surveys were still active, compared with the later tests which were evidently carried out on outcrops or weathered samples, considering the wide variability of the data obtained. Furthermore, data reported on Obi/Lafia coal by Afonja were on seams 12 and 13 which were considered the most economically viable. The relatively high vitrinite reflectance value of Obi/Lafia coal is a strong indication that the deposit was formed at different times and under different conditions compared with the coals of the Anambra Basin. The Adamawa deposit was probably formed around the same period considering the similar physico-chemical characteristics.

It is probable that the deposits of Anambra Basin were formed in-situ considering the low sulphur value. On the contrary, the Obi/Lafia deposit shows many characteristics of oceanic intrusions at the early stages of formation and also during the process of coalification. For example, the sulphur content is high, finely disseminated in the organic matrix, as well as in relatively discrete lumps in form of pyrites. All the coals with the exception of Obi/Lafia (based on representative sample), have significant liptinite content. The maceral is the tar and gas forming petrographic constituent of coal, hence the high gas/tar yield of these coals on carbonization or combustion. The inertinite/mineral matter in all the coal deposits is no more than about 40% which is well within acceptable limits for combustion, and lower than in many coals which are being exploited in many parts of the world.

Table 3.3 Petrographic properties of Nigerian coals.

Coal deposit	Estel (1976) Romax	Afonja (1975, 1979) Vitrinite (%)	Liptinite (%)	Inertinite (%)	Mineral matter (%)	Romax	GCV (MJ/kg)	Uzoegbu et al. (2014) Vitrinite (%)	Liptinite (%)	Inertinite (%)	Mineral matter (%)	Romax	Wuyep & Obaje (2012) Vitrinite (%)	Liptinite (%)	Inertinite (%)	Mineral matter (%)	Romax
Enugu	0.62	61.4	8.4	30.2		0.60	27.6	72-76	6-11	17-18		0.56	44-80	16-6	30-6	10-8	0.54-0.63
Orukpa	3.48	51.4	7.4	41.2		0.5	23.2						91-44	2-36	1-10	6-10	0.45-0.49
Okaba	0.53					0.51	23.9	66	27	7		0.56	40.1	15	35,3	9.6	0.43
Obi/Lafia		79.7	0	20.3		1.0	23.1						41-67	24-5	24-18	11-10	0.2-1.3
Ezimo													69	13	18		0.58
GCV = Gross Calorific Value																	

Inertinite is high in carbon content and is combustible in oxygen-rich atmosphere but has a damping effect on coal thermo-fluidity characteristic on carbonization because it is unreactive. Mineral matter is undesirable in both processes because it forms ash which reduces the useful calorific value and removal may present significant operational problems. Although liptinite and inertinite in coal tend to reduce with maturity due to vitrinization, the proportion in low-grade coals, is determined more by the initial composition of the coal-forming organic matter and the geological history of the deposit. For example, liptinite content may be low because the prevailing conditions favour fast biochemical reactions while high fusinite (inertinite) content may be the result of smoldering fires during peatification under conditions of low aeration.

Vitrinite and liptinite are the only reactive components of coal and these two macerals determine whether or not a coal goes through a plastic phase on carbonization. While liptinite is always reactive, the proportion is low (0-15%) in most coals. In effect, vitrinite reactivity is the main determinant but the degree of reactivity varies with rank and the precursor organic matter. Relatively few coals contain reactive vitrinite and they all belong to the bituminous rank. The fact that Obi/Lafia coal exhibits high plasticity despite the absence of liptinite puts it in the rank of high-volatile bituminous caking coals. Clearly, the deposit is not only older, the geochemical and geophysical conditions of formation and the nature of the original plant debris were different compared with the other proven deposits. In effect, Obi/Lafia coal is potentially suitable as a coking blend constituent. However, the high fluidity needs to be moderated by other low-fluidity coal or inert blend components. Furthermore, the inertinite and mineral matter contents are both syngenetic and epigenetic. This explains the poor response of the coal to cleaning and desulphurization and constitutes a potential constraint to the use of the coal as a metallurgical coking blend component. This is discussed in greater depth in the next chapter.

3.3   PHYSICO-MECHANICAL PROPERTIES OF NIGERIAN COALS

The physical and mechanical properties of coal determine some important economic utilization parameters such as resistance to disintegration during handling, transportation and storage, energy requirements for grinding prior to cleaning or use in a pulverized state, and response to cleaning processes.

3.3.1   Physical Properties of Nigerian coals

Figure 3.6 shows samples of lumps of Nigerian coals. Macro-examination shows that the coals from the Enugu-Okaba belt are black to brownish-black with dark brown streak, mostly dull and lusterless, with thin bands of bright vitrain along fractures at right angles to the bedding. These bright bands are most prominent in Orukpa coal, hence the occasional bright lumps similar to bituminous coal. Okaba coal is dull, friable, and tends to crumble and become unstable when exposed to the atmosphere for a prolonged period. Piles of the coal have been known to ignite on exposure due to exothermic oxidation which generated heat within the pile. The coal decrepitates in storage and calorific value drops. Most of the coals fracture irregularly along the bedding planes and produce a dull gray mottled surface on which fragmentary plant remains are visible (de Swardt, 1963).

Figure 3.6 Coal samples from the major Nigerian coal deposits.
(Jefford, 1963; Afonja, 1979).

3.3.2 Grindability of Nigerian coals

For most applications, coal has to be ground and graded to desired size range which varies depending on application. Most modern coal-fired power generating plants use pulverized coal and grindability has become the coal characteristic for determining the capacity of a pulverizer. There are many methods for measuring this variable but the Hardgrove Grindability Index (HGI) has been adopted by many international standards organizations and is now the standard on which most designs of coal pulverizers are based. The HGI system rates coal hardness from zero to 100. The capacity of coal pulverizers is normally specified as tonnes output when grinding coal with HGI of 50 with a particle size of 70% less than 75 micron and 1 or 2% greater than 300 microns and with a moisture in coal less than 10%. An HGI higher than 50 indicates that a mill processing the coal will have a high output while a low value indicates that the mill will have a reduced capacity. The test is highly non-linear and a change of HGI from 90 to 80 results in a small decrease in mill capacity compared with a change from 50 to 40 which leads to a considerably greater decrease in mill capacity (ACARP, 1998).

Available grindability test results on Nigerian coals are presented in Table 3.4. Both tests were carried out on representative, well preserved samples in British and German coal research laboratories. Although the two tests were carried out using two different standards, the results are comparable. In both tests, Enugu coal was a blend of Onyeama, Okpara and Ribadu mines since the same seam was being mined at three different faces. Enugu and Orukpa coals are very hard coals of similar grinding characteristics. The HGI and specific energy values are higher than typical of coals of the sub-bituminous rank. Okaba coal is considerably softer and the grindability value is typical of low-rank coals and lignites. The unusually low grindability (HGI value) of Enugu and Orukpa coals is due probably to the high inertinite content and this may also explain the unusually high calorific values of both coals since inertinite has the highest carbon content of all coal macerals. Okaba coal, like most lignites is known to be very susceptible to weathering on exposure to oxidative environment. This may to some extent explain the relatively high HGI. Okaba coal is friable especially on exposure to the atmosphere, even for a few weeks. It is therefore of lower value than the other coals for most applications because of its poor handling characteristics and susceptibility to oxidation on exposure and storage. The best option therefore would be to sight a captive power plant close to the mine to reduce the need for holding large stocks. Furthermore, the high HGI would be an asset since mill throughput would be higher. The Zuma power plant is under construction near the deposit.

3.4 CHEMICAL PROPERTIES OF NIGERIAN COALS

The chemical properties of coals tend to vary with maturity, with carbon content, hydrogen and calorific value increasing while volatile matter, oxygen, and water decrease (Figure 3.7). Nitrogen content remains relatively unchanged. Volatile matter can vary significantly in the bituminous rank depending on the nature of the original organic matter and maceral composition, hence the sub-classification into high, medium and low volatile bituminous coals. Ash content depends primarily on the infusible organic and inorganic contents of coal, which in turn depend partly on the amount and nature of inorganic matter deposited with the original coal-forming organic matter as well as subsequent deposits during the coalification process. High inertinite content raises the calorific value since the maceral has the highest carbon content and is combustible.

Table 3.4 Grindability of some Nigerian coals.
(Afonja, 1972; Estel, 1976).

Grindability	Enugu	Orukpa	Okaba
Afonja (1972) (Hardgrove Grindability Index)	39	41.7	71.2
Estel (1976) (Specific work for reduction below 1mm, mkp/kg)	423	405	235

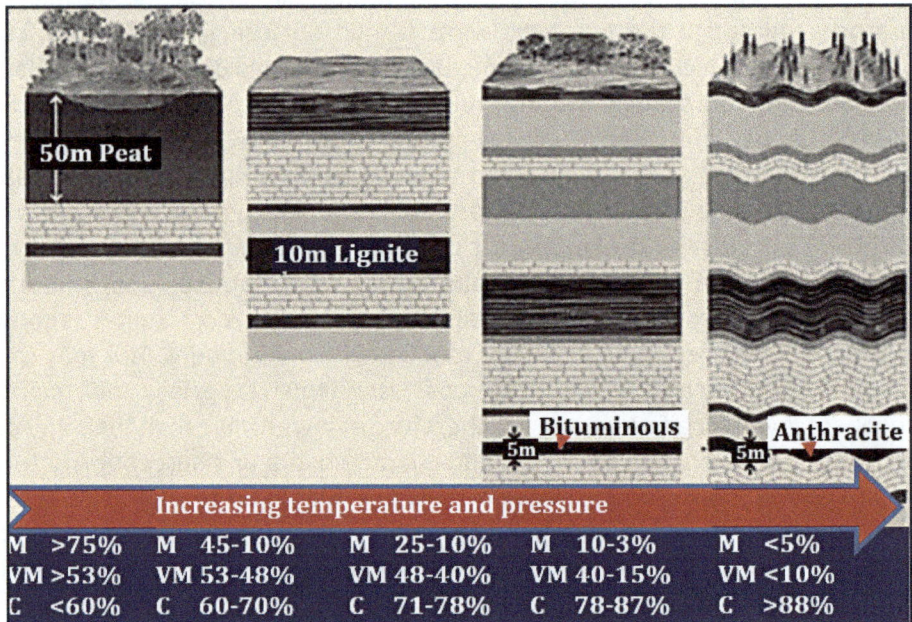

M	>75%	M	45-10%	M	25-10%	M	10-3%	M	<5%
VM	>53%	VM	53-48%	VM	48-40%	VM	40-15%	VM	<10%
C	<60%	C	60-70%	C	71-78%	C	78-87%	C	>88%

Figure 3.7 Changes in chemical composition of coal with maturity. (C= carbon, VM = volatile matter, M = moisture/water. *(Adapted from* adigeologist.blogspot.com/2014*).*

3.4.1   Effect of chemical composition
on the technological value of coal

The carbon and volatile matter contents of coal are the most important chemical variables for assessing its heating value. The moisture content is also important because of the negative effect of high moisture content on heating value of coal. High sulphur content degrades the value of coal significantly because of the environmental implications. Sulphur is also undesirable in coking coal for metallurgical applications. Some chemical compounds present in coal will not burn or contribute to the heating value of coal. They are classified as ash and the proportion in coals is 5-40% or even higher. Many of the compounds undergo endothermic reactions during combustion or carbonization, utilizing a significant amount of the energy potentially available from the carbon content. In effect, the fuel requirement per megawatt of power generation, or per tonne of iron produced in a blast furnace is high compared with low ash coal. The chemical nature of the infusible compounds in coal also determines the physical and chemical characteristics of the ash produced from them.

The fusibility determines the ease with which ash formed can be removed from boiler grates and blast furnaces, and the chemical nature determines the extent of clinkering, fouling and corrosion problems caused in boilers. In blast furnaces a significant proportion of the coke and fluxing compounds charged is used up in fusing and separating the ash from molten iron. Gases such as nitrogen and sulphur compounds evolved from the chemical reactions involving non-combustible matter can have negative effects on the environment. Thus, the economic value of coal depends very much on the proportion and chemical nature of the non-combustible matter classified as ash content as well as the presence of anthropogenic compounds, in particular, sulphur. The quantity, chemical and thermal properties of the ash produced are critical. When heated, coal ash does not melt sharply at any definite temperature but commences to soften at substantially lower temperature than that at which it becomes molten. The nature of this behaviour depends on the chemical composition of the ash. High $Al_2O_3.SiO_2$ content raises the softening temperature; lime and magnesia, ferric oxide, oxides of sodium and potassium lower the fusion temperature markedly. Ferrous oxide produced in the decomposition of ferrous carbonate and sulphate, or by partial oxidation of iron sulphides also lower the fusion temperature. Low ash fusion point leads to the formation of clinker, an undesirable occurrence in coal combustion.

Clinker is a hard mass of refractory particles of ash, fine coal, and fused ash, that clogs a furnace grate, or fluidized bed combustor, reducing the flow of air and often causing furnace or boiler shut down. The formation of clinker is a complex process caused or accelerated by low ash fusion point below about 1300°C. The problem of clinkering can be serious in furnaces that burn low volatile steam coals and anthracites because they require high forced draught and fuel bed temperatures, and use of pre-heated primary air. Usually, such coals are not washed and contain high proportions of iron and sulphur.

3.4.2 Chemical properties of Nigerian coals

Nigerian coals fall mainly in the lignite and sub-bituminous classes, with the exception of Obi/Lafia coal which exhibits some characteristics of high-volatile caking bituminous coals. The main chemical properties of Nigerian coals are summarized in Table 3.5a. The data was compiled from various studies (Eisen, 1962; Arthur D. Little 1965; Afonja 1974, 1975, 1979; Estel, 1976;). Enugu coal was mined at three different faces: Onyeama, Okpara and Ribadu mines. Various analyses of coal samples from the three mines have shown no significant difference. It should be noted however that much of the data presented in Table 3.5(a) is unreliable since many of the tests were conducted on borehole samples, samples stored over long periods or outcrops. In view of the high heterogeneity of coal and high weatherability of low-grade coals, many of the coals samples tested were either non-representative of the deposits, or highly oxidized, or both. Furthermore, the basis of analysis for determining the proximate, ultimate and calorific values was not stated in some of the investigations. For example, the value of a chemical composition variable can vary by up to 30% depending on whether it was expressed on as received (ar); dry; dry, ash-free (daf); dry-mineral matter-free (dmmf) basis (3.5(b).

In recent years, many investigations have been carried out on the Obi/Lafia, Adamawa and Gombe coal deposits. None of these deposits have been mined, hence the tests were done on outcrops. Some of the results are summarized in Table 3.5c. One important observation from the chemical composition of Nigerian coals is that moisture and carbon contents decrease from the south (Enugu) to the north (Okaba) while oxygen content increases in the same direction. The results of petrographic reflectance measurements discussed above also show a decrease from south to north, indicating that maturity of the coals decreases from south to north. However the conclusion that maturity of Nigerian coals decreases from south to north does not apply to Obi/Lafia and possibly Adamawa coals which are further north but are or higher rank, or the Gombe coal located even further north but is lignitic. The only logical explanation for these anomalies is that the northernmost two coal deposits were established from organic matter of different chemical nature under different geological timelines and conditions, different history of earth movement   and marine incursions, etc.

Table 3.5a Chemical composition of Nigerian coals, (daf = dry, ash-free).
(Arthur D. Little, 1965; Afonja, 1972, 1974, 1979; Eisen, 1976).

Quality Parameter	Enugu	Inyi	Orukpa	Okaba	Obi/Lafia (washed)	Ogwashi Lignite
Moisture (%)	8	12.1	12.5	9.5		
Ash (%, dry)	3.5-21.6	6.5	8-9	9.6-15.9	14.7	14.4
Volatile matter (%, daf)	46.7	-	51.9	43.9	31.7	64.5
Fixed carbon (%, dry)	43.1	-	48.7	50.6	49.6	-
Carbon (%, dry)	79.3	-	73.9	78	80.4	-
Hydrogen (%, dry)	5.1	-	4.6	6.1	5.6	6.6
Total sulphur (%, dry)	0.75	1.47	0.46	0.7	1.5-6.5	0.78
Sulphate sulphur (%, dry)	0.01	0.02	0.02	-	0.02	0.05
Pyritic sulphur	0.05	0.75	0.02	-	1.4-6.5	0.02
Organic sulphur (%, dry)	0.69	0.67	0.43	-	0.07-0.3	0.71
Gross calorific value (MJ/kg, daf)	33.8	24.3	31.4	30.7	32.3	30.5

Table 3.5b Effect of basis of analysis on values of coal properties. [AR=as received; MMF=moist, mineral matter-free; DMMF= dry, mineral matter-free]. *(Afonja and Olofinjana, 1988).*

COAL PROPERTY	COAL 1			COAL 2			COAL 3			COAL 4			COAL 5		
	AR	MMF	DMMF	AR	MMF	DMMF	AR	MMF	DMMF	AR	MMF	DMMF	AR	MMF	DMMF
Moisture %	15.9	20.3	N/A	4.7	7.1	N/A	13.5	15.4	N/A	7.1	8.1	N/A	2.4	3.2	N/A
Ash %	20.3	N/A	N/A	32.3	N/A	N/A	11.8	N/A	N/A	12.5	N/A	N/A	24.7	N/A	N/A
Volatile Matter %	25.0	32.0	40.3	30.1	46.3	49.9	32.5	37.3	44.1	40.6	47.0	51.2	22.8	31.1	32.2
Fixed carbon %	38.8	49.8	62.5	32.9	50.6	54.6	42.2	48.4	57.4	39.8	46.1	50.2	50.1	68.5	70.3
Carbon %	46.8	60.0	75.4	48.6	74.9	80.7	58.5	67.2	79.5	62.7	72.7	79.2	60.8	83.2	86.0
Gross calorific value MJ/kg	18.1	23.1	29.1	20.4	31.4	33.8	24.5	28.1	33.3	26.3	30.4	33.2	24.8	33.9	35.1

Table 3.5c Chemical properties of Obi/Lafia, Adamawa and Gombe coal deposits.

Coal Property	Akpabio et al., (2008)			Ajani & Makoyo, (2008)	Adeleke, (2005)	
	Obi/Lafia	Doho	Lamja	Chikila	Chikila	Lamja
Moisture (wt %)	2.91	3.90	3.08	8.9	5.2	1.9
Ash (%, db)	20.66	29.91	11.87	21.3	9.6	11
Volatile matter (%, db)	27.29	43.44	40.01	36.3	48.8	43.5
Fixed carbon (%, db)	46.23	21.98	44.23	34.58	51.2	56.5
Total sulphur (%)	2.91	0.77	0.81	0.34		
Gray King Assay	G3	A	B	A		
Free Swelling Index	6	0	0		1.0	1.5
Dilation (%)	23	-5	-5			
Fluidity (ddpm)	21	0	0			

3.4.2.1 *Ash content of Nigerian coals*

Ash content values quoted for Enugu coals in literature range from 4 to 26%. The coal is drift-mined and samples taken from mine cars at different times by the same investigator had yielded ash values between 5% and 22%. The wide range of values is partly a result of inappropriate sampling. In many cases core samples were analyzed, in others they were handpicked. In either case, the sample cannot be representative of a coal deposit, especially when the seams have band intrusions as in the Enugu deposits. Investigations have shown that Enugu coal is mined along with considerable amount of shale which looks very similar to coal but is softer, lighter and of different chemical composition. This is evident from the result of float-sink tests in Table 3.6. The ash content of washed Enugu coal is only 9-10%. On the basis of this ash value, the volatile content is only about 41%.

Clay mineral matter is predominant in Enugu, Orukpa and Okaba coal deposits but the structure of Obi/Lafia deposit is different. Apart from clay minerals, the seams are also interspersed with and intruded by iron pyrites and the morphology of the mineral matter in the coal matrix differs significantly. Clay minerals are present in Enugu coal in isolated grains but are intimately intergrown with the organic matter in Orukpa, Okaba and Obi/Lafia coals. Iron pyrites is also finely dispersed in Obi/Lafia coal matrix. Orukpa and Okaba deposits are surface-mined and therefore contain relatively lower ash because the non-carbonaceous overburden is first removed before the coal is mined and it is relatively easy to discriminate between coal and shale. Washing makes little difference to ash content, hence washing is not necessary. The quality of all the coals makes them potentially suitable for steam raising and power generation.

Table 3.6 Response of some Nigerian coals to float-sink tests. *(Estel, 1976).*

Coal/size range (mm)	Raw coal		Washed coal (sg=1.5g/cm³)		Total yield
	Content (%)	Ash (%, dry)	Yield (%)	Ash (%, dry)	(% of raw coal)
Enugu					
50 - 16	52.1	20.7	81.8	9.6	74.4
16 – 1	39.7	20.5	80.5	9.5	
> 1	8.2	30.3			
Orukpa					
50 – 16	80.0	6.5	99.7	6.5	95.7
16 - 1	17.1	14.3	0.3		
> 1	2.9	24.8			
Okaba					
50 - 16	77.9	10.1	99.7	10.1	96.6
16 - 1	19.1	9.0			
> 1	3.0	12.2			

High ash content reduces the value of any coal irrespective of potential application, but can often be reduced by cleaning processes depending on the nature and morphology of the ash-forming components of the coal. They are easy to remove when they are in relatively discrete lumps but when finely disseminated in the coal matrix, ash reduction can be difficult and expensive. Prior to closure, Enugu coal was mined by manually fracturing the coal seam using hydraulic and pneumatic hammers and the products were used without washing. Mechanization of the mine would produce coal of significantly higher ash content because non-carbonaceous material above, below and embedded in seams would be mined along with coal. In effect, the coal must be cleaned.

In a recent study of the response of Obi/Lafia coal to cleaning, Jatau et al., (2013) carried out float and sink and froth flotation tests on small samples. Three organic chemicals were used in the float and sink tests and an ash content of 5.7% at 80% yield was obtained with a medium of specific gravity of 1.6. Froth flotation at Ph 7 yielded coal of ash content of 6.6%. It should be noted however that the tests were carried out on 500-gm samples in small laboratory beakers, using expensive chemicals. The samples tested were hand-picked from outcrops. Considering the wide variability In chemical properties between seams in the deposit, not much inference can be drawn from these results.

Coal ash chemistry is an important parameter in evaluating the quality of coal for combustion or carbonization. Reactions involving some ash-forming inorganic matter are endothermic and reduce the useful heat content of coal. Also, ash that cakes creates problems in the combustion furnace. Two important quality indices for coking coals and coal blends are acidity ratio (AR) and mean basicity ratio (MBI), determined from equations 3.1 and 3.2 respectively.

$$AR = \frac{\% Al_2O_3 + \% SiO_2}{\% Fe_2O_3 + CaO + MgO} \qquad (3.1)$$

$$MBI = \frac{100 \times \% Ash \times Na_2O + CaO + MgO + Fe_2O_3]}{[(100 - VM) \times (SiO_2 + Al_2O_3)]} \qquad (3.2)$$

The presence of ash-forming mostly inorganic compounds is determined by the geochemical and geological history of a coal deposit, in particular, the original pit forming matter, the geology of the location, and oceanic transgression and regression which determine the type and quantity of non-organic matter deposited with or on top of the coal-forming matter over millions of years. There is little information about the chemical composition of the ash in Nigerian coals. The available data is presented in Table 3.7. Silica is the predominant compound. The significant disparity in alumina content probably arose from differences in sampling. The sample analyzed by Afonja (1976) was taken and prepared in accordance with BS 1016 specifications for coal analysis, from a 100kg sample collected from ten wagons of freshly mined coal at Onyeama mine. There is no information on the sampling used by Mayaleeke et. al., or from which of the four mines in Enugu zone the sample was collected. However, the tests were carried out many years after the mines were closed and likely came from outcrops or exposed coal piles at the mine.

Table 3.7 Ash Chemistry of Enugu coal.

Chemical compound (%)	Afonja (1976)	Mayaleeke et al., (2009)
SiO_2	64.6	61.2
Al_2O_3	25.9	7.1
Fe_2O_3	3.6	2.9
CaO	1.1	1.4
K_2O	0.6	0.2
TiO_2	2.2	
MgO		0.2
Acidity Ratio (AR)		15.2
Mean Basic Index (MBI)		1.1
Total Alkali		0.4

3.4.2.2 *Heating value of Nigerian coals*

The heating value of coal, also known as calorific value is the amount of useful energy that can be obtained per unit weight of coal. This excludes the energy required for moisture removal, energy required for the endothermic chemical reactions during combustion or carbonization, and the energy required for fusing the ash content. In effect, the useful energy content increases with increase in the combustible contents (carbon and hydrogen) and reduces with moisture and ash content. Calorific values for all Nigerian coals fall between 24 and 34MJ/kg (dry, ash-free) (Table 3.5a). These values are significantly higher than is usual for low-grade coals. As mentioned earlier, this is due probably to the high inertinite content of all the coals. Inertinite has the highest carbon content of all macerals in coal and carbon is the main source of energy. Using the ASTM system which uses only calorific value and caking behaviour to classify low-grade coals, Enugu, Orukpa and Okaba coals would fall in the rank of high-volatile bituminous non-caking coals while Ogwashi coal would be classified as sub-bituminous coal. The high heating values and non-caking behavior of all Nigerian coals make them potentially very suitable for steam raising and power generation.

3.4.2.3 *Sulphur content of Nigerian coals*

Sulphur is considered one of the major impurities in coal ranging from 0.1 to about 10% or higher depending on the structure of the coal deposit, and low levels are required for most applications. On carbonization or combustion of coal, sulphur dioxide is released into the atmosphere, causing environmental damage. In steelmaking, sulphur passes into the steel and causes embrittlement. Also, exothermic chemical reactions involving sulphur compounds can significantly increase the self-ignition propensity of coal, in particular, low grade coals in storage. Sulphur occurs in coals in four forms: as pyritic, organic, sulphate or elemental sulphur. Pyritic and organic forms together account for the large majority of sulphur in coal (Morrison, 1981). Sulphate sulphur occurs mainly as gypsum and iron sulphate, the latter being mainly a product of oxidation of pyrites on prolonged exposure to the atmosphere. The sulphur content of coal has become a major variable in assessing the economic value of a given coal deposit. Ferrous sulphide is the most common form of sulphur occurrence in coal and it occurs mainly in two forms: pyrites which has cubic structure and marcasite which has orthorhombic structure. Both are usually classified as pyrites since it is the more common form. Pyrites occurs in coal in narrow seams or veins up to 150mm thick and several hundred millimetres long, as nodules mostly 10-20µm in diameter, or as discrete crystals predominantly 1 – 40µm. Organic sulphur in coal is usually very small and is believed to originate from the plant and animal debris that formed the peat from which the coal evolved. Very little is known about its presence in coal, how its quantity is determined, or how it can be removed. Sulphur rarely occurs in coal in elemental form, and it is found mainly in weathered coals. All Nigerian coals with the exception of Obi/Lafia coal have low sulphur content as shown in Table 3.8 (Afonja, 1975).

The sulphur in Enugu coal is predominantly organic and further reduction would be difficult and expensive. In any case, it is within acceptable limits for most uses of coal. The sulphur content in Obi/Lafia coal is not only high but it varies widely from 1 to 11%. This is due apparently to the structure of the coal bed which features randomly distributed sulphur compounds in layers adjacent to coal strata. The sulphur is 80-85% pyritic, some in discrete lumps (Figure 3.8) but much of it is finely disseminated in the coal matrix structure and would be difficult to remove by typical cleaning methods. All the tests that have been carried out so far on Obi/Lafia coal were on borehole samples, hence it has not been possible to obtain representative samples of the deposit. Therefore data on the deposit should be regarded as preliminary.

Table 3.8 Sulphur in Nigerian coals. *(Afonja, 1975).*

Sulphr type (%, dry)	Enugu	Inyi	Orukpa	Okaba	Obi/Lafia	Ogwashi
Total sulphur	0.75	1.47	0.46	0.7	1.53-11.1	0.78
Sulphate sulphur	0.01	0.02	0.02	-	0.01-0.21	0.05
Pyritic sulphur	0.05	0.75	0.02	-	1.39-6.52	0.02
Organic sulphur	0.69	0.67	0.43	-	0.07-0.29	0.71

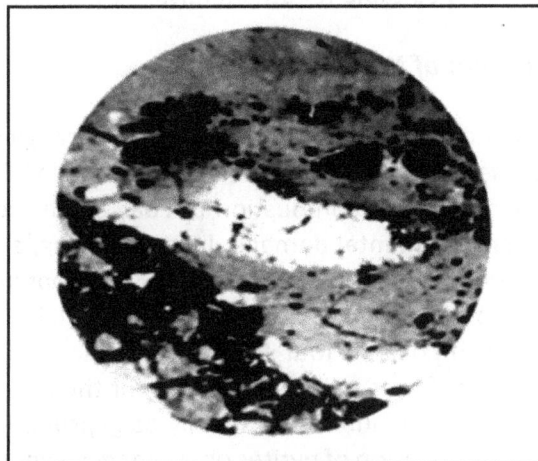

Figure 3.8 Sulphur (white, in form of pyrites) in Obi/Obi/Lafia coal. *(Afonja, 1979).*

3.4.2.4 *Volatile matter content of Nigerian coals*

There is an established relationship between coal rank and volatile matter content. The higher the rank, the lower the volatile matter (Figure 3.7). Also, high volatile matter is indicative of low carbon content. All Nigerian coals with the exception of Obi/Lafia and possibly Adamawa coals have volatile matter higher than 40% (dry, ash-free). The high volatile matter content and low carbon values indicate that the coals are relatively young and of the sub-bituminous to high-volatile bituminous rank. The high volatile matter content also indicates that the coals are unlikely to have any appreciable coking characteristics. Coking coals normally have a volatile matter content of between 20 and 32%. The high value of volatile matter is however an advantage if the coal is for use as fuel since most of the volatile matter is combustible. High volatile coals also have a high potential as feedstock for the production of chemicals. On the basis of the carbon content, all the coal deposits, proven or identified, are of the sub-bituminous rank with the exception of Ogwashi deposit which is lignitic and Obi/Lafia coal which is in the high-volatile bituminous rank.

3.4.3 *Other coal resources of Nigeria*

Coal deposits are classified as resources until the extent and quality of the deposit have been established. Apart from the deposits discussed above, there are many others that have not been characterized. The Adamawa coal deposit is of particular interest because of its possible caking properties. Although it had been known for a long time that there are coal occurrences in Adamawa State of Nigeria, not much attention had been given until recently. The deposit outcrops in Chikila and Lamja but the Lamja seam is believed to be too thin for commercial exploitation. Another deposit in Gombe (Doha) is lignitic, based on scanty, available data (Table 3.5c). The deposit is being mined locally for cement production and a coal-fired power plant is also planned. It should be noted that all available data on chemical and thermoplastic properties of these coals were carried out on small samples picked from outcrops. Apart from the fact that they are not representative of the deposits, they are probably heavily weathered, hence the results should be taken as indicative only.

REFERENCES

Adeleke, A. O., Ibitoye, S. A., Afonja, A. A. and M. M. Chagga (2011) "Multistage caustic de-ashing of Nigerian Obi-Lafia coal." *Petroleum & Coal,* Vol. 54, No.4, pp. 259-265

Adeleke, A. O. (2005). *Studies on blend design for enhancing the cokeability of Lamza coal.* M.Sc. Thesis, Department of Metallurgical and Materials Engineering, Federal University of Technology, Akure, Nigeria.

Afonja, A. A. (1974). Evaluation of the coking characteristics of some Nigerian coals by thermogravimetry. West African Journal of Science, 19(2), Pp. 171-174.

Afonja, A. A. (1975). "Chemical, Petrographic and Coking Studies of Enugu Coal." Nigerian Mining and Geology, 1975, 12, p.40.

Afonja, A. A. (1976). "Petrography of Nigerian Coals." Research Report No. IFE/CHE/CP/05, Coal Research Laboratory, Ife.

Afonja A. A. (1976). " Deashing and desulphurization of Nigerian coals. Research Report No. IFE/CHE/CP 06.

Afonja, A. A. (1979). "Further Studies on the Obi/Lafia Coal Deposit (Phase II)." Research Report No. IFE/CHE/CP/11, Coal Research Laboratory, Ife.

Ajani, O. O. and M. Makoyo (2008). "Assessment of Chikila coal for use as metallurgical coke." *JOMAR*, 5(1&2, Pp. 61-65.

Akpabio, I. O., Chagga, M. I., and A. Jauro (2008). "Assessment of some Nigerian coals for metallurgical application." *Journal of Minerals and Materials Characterization & Engineering*, Vol. 7, No. 4, Pp. 301-306.

Arthur D. Little Inc. (1965). Revitilizing the Enugu Coal Industry Through Large-Scale Carbonization. Duffryn Technical Services Ltd. Chemical Utilization Project Report of the Under Secretary of State for the colonies, London.

Bustin, R. M., A. R Cameron, D.A. Grieve, and W. D. Kalkreuth (1983). *Coal Petrology, its principles, methods and applications. 2nd edition: Geological Association of Canada Short Course Notes 3, 230p.*

Crellings (2016). Crellings petrographic atlas of coals, www.coalandcarbonatlas.siu.edu/coal macerals). Accessed, July, 2016.

de Swardt, A. M. J. and O. P. Casey (1963). "The coal resources of Nigeria." *Geological Survey of Nigeria* Report No. 28.

Diesel, C. F. K. (1992). *Coal bearing depositional systems*. Springer Verlag, Berlin.

Estel (1976). "Investigations into possibilities of preparing and carbonizing Nigerian coals from Enugu District." Estel Exploration und Bergbau GmbH, Dusseldorf.

Eisenbau-Essen (1962) Carbonization of Enugu Coal. Report to the Eastern Nigeria Government.

Jatau, B. S., Amoka, I. S. and S. I. Fadele (2012). "Washability and froth flotation tests of Lafia-Obi coal deposit, Nasarawa State, North-Central Nigeria."International Journal of Natural Sciences, Vol. 1, No. 1, Pp. 1-9.

Jefford, G. (1963). "The Physical and chemical properties of Nigerian coals." *Geological Survey of Nigeria* Report No. 28.

Mayaleeke, A. H., Adeleke, A. O., and D. A. Dashak (2009). "Studies on the ash chemistry of the Nigerian Enugu coal a a blend component in metallurgical cokemaking." *The Pacific Journal of Science and Technology*, Volume 10, No. 2, Pp. 782-787.

Morrison, G. F. (1981). *Chemical desulphurisation of coal*. ICTIS/TR15; London, U.K., IEA Coal Research; 72pp.

ROE GMBH (1982). "Feasibility Study of the Carbonization of Okaba Coal." Final Report prepared for the Nigerian Coal Corporation. Berlin, West Germany.

Simpson, A. (1954). "The geology of part of Onitsha, Owerri and Benue Province. The Nigerian Coal field." Geology Survey Nigeria Bull. 24.

Stach, E.M. Mackowsky, M. Teichmuller, G.H. Taylor, D. Chandra and R. Teichmuller (1982). *Xtach's textbook of coal petrology; Stuttgart¨Gebruder Born-traeger.*

Uzoegbu, U. M., Ekeleme, I. A., and U. A. Uchebo (2014). "Oil Generation Capacity of Maastrychtian Coals from the Anambra Basin, SE Nigeria." *The International Journal of Engineering and Science*, Vol. 3, Issue 4, Pp. 33-46.

Wuyep, E. O. and Obaje, N. G. (2012). "Applications of some coal deposits in the Anambra Basin and Middle Benue Trough of Nigeria." *Journal of Earth Science and Engineering*, Vol. 2, Pp. 220-234.

4 Thermochemical and weathering properties of Nigerian coals

4.1 INTRODUCTION

The behavior of coal on heating depends on two major variables: coal rank and coal type. For example the heating value of coal is determined primarily by the carbon content but also by the volatile matter and moisture contents. The amount and chemical nature of mineral matter content also reduce calorific value since many minerals decompose endothermally and may use part of the available thermal energy. Thermoplasticity (ability to go through a softening phase on heating) is mainly a function of the coal rank and only some coals in the bituminous rank exhibit any significant plasticity on heating. Coal type (nature and relative proportions of the three maceral groups) and mineral matter content also have a significant influence on thermoplasticity. Only reactive macerals have positive influence while unreactive components and mineral matter may act as diluents. The relative proportion of the mineral matter content of coal compared with organic constituents also determines the amount of ash formed on carbonization and combustion. Ash has a negative value in coal. Apart from being a diluent, it has to be removed from combustion grates or by slagging in ironmaking, hence the chemical composition and fusibility are important.

4.2 COMBUSTION AND CARBONIZATION PROPERTIES OF NIGERIAN COALS

The complex processes that take place when coal is pyrolyzed in surplus air (combustion) or minimum air (carbonization are well treated in many books (Speight, 2012, Afonja, 2017). The important variables are physical characteristics, heating value, plastic behavior, and ash characteristics. The physico-chemical characteristics of Nigerian coals have been discussed in Chapter 3.

4.2.1 Combustion characteristics of Nigerian coals

All Nigerian coals including the lignite deposits have good heating value, considered to be high compared with typical low-grade coals. This is due to the relatively high inertinite contents. Inertinite has the highest carbon content of all coal macerals and, can burn to produce high energy if there is sufficient oxygen available. Modern coal-fired plants supply oxygen-rich air to burners to ensure complete combustion of inertinite. Unburnt inertinite ends up as ash and removal from combustors can cause serious operational problems depending on the fusibility characteristics. The ash levels of Okaba and Orukpa deposits which are mined by opencast technique fall within acceptable limits for combustion, even without cleaning. Also, the sulphur contents are low and no pre- or post combustion desulphurization may be necessary. The coal deposits in Enugu area which are only recoverable by underground mining technique contain high gangue and washing is required to bring the ash content to acceptable levels. However, they have good cleaning characteristics (See Chapter 3). Also, if the deposits supply local captive power generating plants, washing may not be necessary. Furthermore, most modern power plants now deploy fluidized-bed combustors which can process many types of unwashed, pulverized coals.

Nigerian coals have no thermoplastic properties (they are thermally unreactive), the exception being the Obi/Lafia coal which is unsuitable for combustion because of its high sulphur content among other problems. There are indications that the Adamawa deposit may also have some caking characteristics. Low or non thermoplasticity is considered a positive characteristic in coal combustion because caking coals tend to agglomerate in pulverized coal combustors and cause operational problems. Such coals are often thermo-treated to remove the plasticity prior to combustion.

Modern power generating plants use pulverized coal ground to minus 3mm for steam generation, mainly in fluidized-bed combustors. Most Nigerian coals are hard to grind and so far have been used only in lump form on combustion grates. Coal-fired power plants which are currently planned will feature fluidized-beds and the coal needs to be pulverized. However, pulverization should not be a problem, except that higher energy will be required and throughput will be lower. All the deposits that are currently being processed for power generation have low contents of sulphur and other anthropogenic compounds. Furthermore, fluidized-bed combustors facilitate relatively easy removal of pollution products of coal combustion.

4.2.2   Basic theory of coal carbonization

Coal is heated in minimum air (carbonized) to expel the volatile matter content without burning the coal. The residue is nearly pure carbon (over 98%) with very high heating value. Carbonized coal (coke) has potential utility in many applications including iron and steel production, smokeless fuel briquettes, activated carbon, graphite, industrial diamond, etc. Depending on the thermochemical properties of a particular coal, the coke may be in a granular, or solid forms of varying strength and porosity. The behavior of coal in the temperature range of around 350 to 750°C to a large extent determines the potential areas of utilization. At around 350-380°C virtually all coals contract although the temperature at which contraction starts and the degree of contraction vary depending on coal rank.

Coal contraction on heating is due to the expulsion of some volatile matter, mostly moisture, saturated paraffin hydrocarbons, and small quantities of unsaturated hydrocarbons, hydrogen, carbon monoxide, carbon dioxide and hydrogen sulphide. Some coals start to soften from the outermost layers at around 375 to 400°C, forming a thin, plastic envelope (Figure 4.1, stage 2). This phase is also known as *mesophase.* As temperature increases, the thin plastic layer extends towards the centre and more gases are produced.

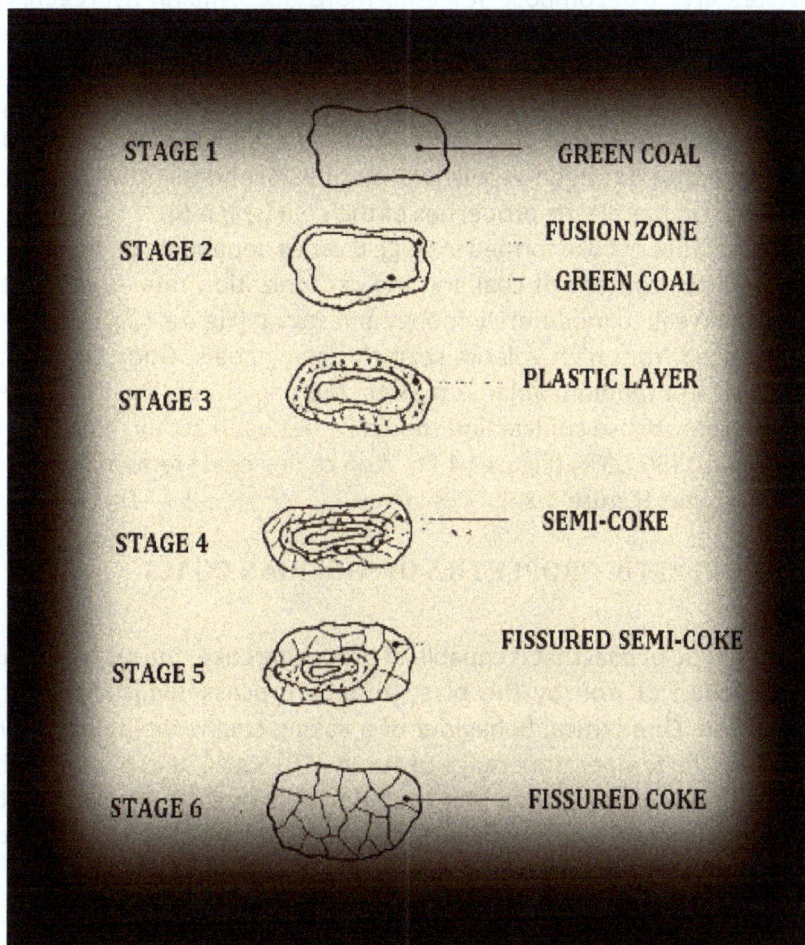

Figure 4.1 Stages of transformation from coal to coke during carbonization. *(van Krevelen, 1961, Lowry, 1963).*

The internal pressure increases due to the restriction imposed on the expulsion of the gases by the plastic layer and the coal starts to swell (stage 3).This stage is absent in non-caking coals. The extent of expansion of coal in the mesophase depends on the thickness of the plastic layer, the rheological properties of the mesophase, and the amount of volatile matter released, all of which vary between coals, even in the same rank. Coals which do not form a plastic phase do not dilate, only contract due to the expulsion of gases. At around 550°C about 70 to 75% of the gases produced have escaped (some of the gases also decompose) and the coal starts to reconsolidate to form a porous, reactive black solid mass often referred to as *semicoke*. The rheological behaviour of coal in the 350-550°C temperature zone determines its ability to form coherent coke. Thermoplasticity in coals ranges from none to very high and more than 80% of the global coal resources have little or no thermoplastic properties. Coals with medium plasticity form the strongest semicoke, highly thermoplastic coals form weak, fragile semicoke while those which have no thermoplastic properties remain in granular or pulverized form. Between 550 and 750°C the fusion zone has moved right to the centre and the semicoke starts to decompose, releasing more gases, mainly hydrogen, carbon monoxide, carbon dioxide and higher hydrocarbons. A steep temperature gradient develops in the coal mass due to relative differences in the reaction profiles from the coal mass surface to the centre. The thermal gradient combined with the pressure created by the secondary degasification cause the semicoke to crack and fissure (stage 5). At around 900 to 950°C nearly all the volatile matter has been expelled (around 98%) and the residue is fissured or granular coke depending on the plastic properties of the coal (stage 6).

The structure of coke formed in stage 6 varies depending on the rheological properties and volatile matter content of coal and the carbonization rate. The structure ranges from very spongy and weak to moderately spongy and strong (Figure 4.2). Cokes (a) to (c) are typical of cokes produced from high volatile strongly caking coals. Good coking coals should have optimal fluidity and medium volatile matter. Dilation should be moderate (50–100%) and optimum post mesophase contraction should be between 10 and 20%. The coke produced is strong, yet porous (50-55%) (Figures 4.2f). Non caking coals remain granular throughout the carbonization stages (Figure 4.2g).

4.3 THERMOPLASTIC PROPERTIES OF NIGERIAN COALS

Coking coal is a type of coal that is capable of producing coke, the most expensive raw material for the production of iron by the blast furnace process which is still the major global production route. One critical behaviour of a caking coal is the ability to soften, degas and reconsolidate in the temperature range of about 350–550°C. In the process the coal contracts, dilates and ultimately transforms into the familiar spongy semicoke structure. All coals go through the contraction phase which occurs due to devolatilization, but only caking coals show any subsequent dilation.

Figure 4.2 Internal structures of different cokes, (a)-(e) from highly volatile caking coals; (f) from medium volatile coking coals; (g) from non-caking coals. *(Partly from van Krevelen, 1961, Lowry, 1963).*

Coals which have high carbon content, moderate volatile matter content, moderate dilation and fluidity properties, and as little infusible materials as possible are known as *prime coking coals* and tend to form strong, porous coke of metallurgical quality, with optimum reactivity to carbon monoxide and water vapour, a porous structure and sufficient strength and abrasion resistance to resist breakage during handling, charging and passage through the blast furnace. Relatively few coals fall in the category of prime coking coals, most caking coals have either inadequate or excessive properties. This explains why most metallurgical coke ovens use blends of several coals with extreme properties to achieve thermochemical properties as close as possible to prime coking coals.

None of the Nigerian coals with the exception of Obi/Lafia coal has any appreciable caking properties. They are of the sub-bituminous/lignite ranks and no coal in this group has been found to have caking properties. However, carbonization tests have shown that Enugu coal has slight fluidity in the mesophase region, evident from post-fluidity grain coalescence (Table 4.1). An increase in the +20-16 mm size proportion from 18.1% to 28.5% shows that Enugu coal has some fluidity in the 350-550°C temperature range and there is some post-fluidity grain fusion. On the contrary, Orukpa and Okaba coals suffer grain degradation as a result of carbonization. Obi/Lafia coal is marginal-bituminous in rank, it goes through contraction, dilation and plastic transformation and exhibits caking properties. However, the volatile matter is excessive and this is probably responsible for the friable coke produced from the coal. In the devolatilization temperature range the gases force their way from inside through the thick plastic shell, causing numerous blow holes, resulting in weak, fragile coke. This behaviour is typical of high volatile bituminous caking coals.

4.3.1   Dilation and fluidity properties of Nigerian coals

All Nigerian coals with the exception of Obi/Lafia coal do not soften appreciably on heating and show only contraction but no dilation as shown in Table 4.2 and Figure 4.3. The fluidity curves are shown in Figure 4.4. A foreign high grade coking coal has been included for comparison. Unwashed Obi/Lafia coal has moderate caking properties but washed coal has very high thermoplastic properties as confirmed by the dilation and fluidity curves. The Gray-King Coke value of G9 also supports this conclusion. However, the yield is low and, although it can be blended with other coals with low fluidity, the sulphur content of the cleaned coal is still too high for metallurgical applications. In spite of the problem of high sulphur content, the coal is potentially suitable for power generation. Most modern power plants feature fluidized-bed combustors and most of the sulphur can be removed by addition of lime. Also, the coal would need to be pre-oxidized to remove the plastic properties and avoid coagulation in the bed.

Table 4.1 Changes in grain size distribution of some Nigerian coals on carbonization at 550°C. *(Afonja, 1972).*

Grain size (mm)	Enugu		Orukpa		Okaba	
	Fresh	Calcined	Fresh	Calcined	Fresh	Calcined
20-16	18.7	28.5	11.1	3.0	3.2	1.1
16-1	75.5	65.0	71.8	74.1	73.3	67.8
> 1	5.8	6.5	17.1	22.9	23.5	31.1
	100	100	100	100	100	100

Table 4.2 Thermoplastic and caking properties of Nigerian coals, and a prime coking coal. *(Afonja, 1972, 1974, 1979; Eisen, 1976). *See Table 3.5c for sources of data on Adamawa coal).*
Note: values of chemical composition are averages of all available data.

Quality Parameter	Enugu	Inyi	Orukpa	Okaba	Obi/Lafia (washed)	Ogwashi Lignite	*Adamawa Lamza	*Adamawa Chikila	Prime coking coal
Moisture (%)	3.5		8.3	9.6					
Ash (%, dry)	21.6		8.0	10.0					
Volatile matter (%, daf)	46.4	42.6	47.1	43.8	31.7		43.5	48.8	15-22
Fixed carbon (%, daf)	54.8		52.9	56.2					
Carbon (%, daf)	82.1		77.7	81.4	88.4				85-89
Hydrogen (%, daf)	5.84		5.38	6.45					
Nitrogen (%, daf)	1.88		1.98	1.81					
Sulphur, total %, (dry)	0.79		0.55	0.60					
Oxygen (%, daf)	5.8		14.3	13.4					
Gross calorific value (mJ/kg, daf)	33.8	24.3	31.4	30.7	32.3	30.5			33-36
Free Swelling Index (FSI)	1	½	0	½	8-9	0	1.5	1.0	8
Dilatometer softening point (°C)	356	357	357	375	357				380-385
" Reconsolidation point (°C)	477	483	462	426	468				470-475
" Softening range (°C)	121	126	105	51	111				90-95
" Contraction (%)	27	15	18	30	28				10-25
" Dilation (%)	0	0	0	0	190				120-180
Gray King Coke Type	A				G9		A	A	G8
Gieseler fluidity (ddpm)	3				27,600				900-1200
Mean maximum reflectance (%)	0.45/0.62		0.53	0.48	0.95				1.1-1.2
Vitrinite (%)	61.4				81.7				82-83
Exinite (%)	8.4				0				5-6
Inertinite (%)	30.2				12.1				11-12

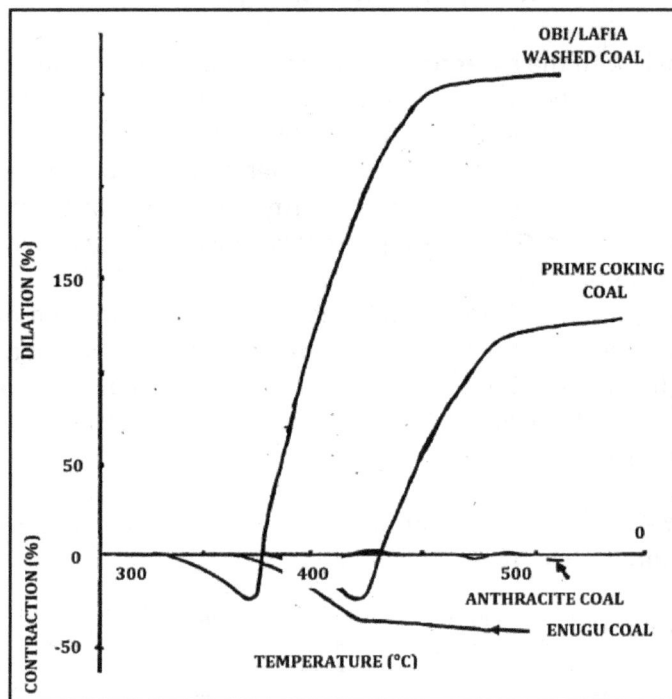

Figure 4.3 Dilation curves for Nigerian coals and typical coking coals. *(Afonja, 1974).*

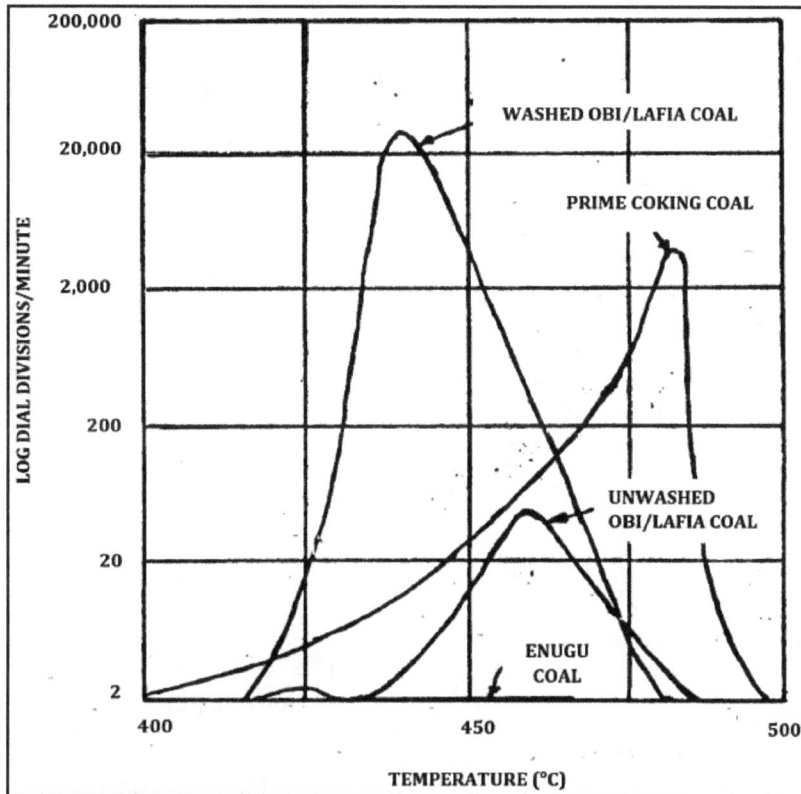

Figure 4.4 Fluidity curves of some Nigerian and some typical coking coals. *(Afonja, 1979).*

The fluidity curves in Figure 4.4 confirm the results of the dilation tests. Enugu coal has no fluidity, Obi/Lafia coal has moderate fluidity but the washed coal has excessive fluidity compared with a prime coking coal. The coke produced is friable and crumbles easily. However, the coal can be blended with several other coals of lower rheological properties for metallurgical coking. Also, addition of coke breeze prepared from Enugu coal moderated the dilation considerably and the product coke was much stronger. In fact, coke of metallurgical quality could be produced with about 30% coke breeze. This is discussed in detail in a later chapter.

4.3.2 Thermogravimetric properties of Nigerian coals

Some Nigerian and several foreign coals have been studied by thermogravimetry to determine their devolatilization characteristics and the results are shown in Figures 4.5 to 4.8 (Afonja, 1974, 1979). The thermograms gave volatile matter values which were in good agreement with values obtained by proximate analysis and confirm that Enugu, Okaba and Obi/Lafia coals are low-rank, high volatile coals. The highest rank coals have the lowest rate of weight loss. Also, an approximately linear relationship was established between the maximum rate of devolatilization and rank as established by volatile matter content (Figure 4.9).

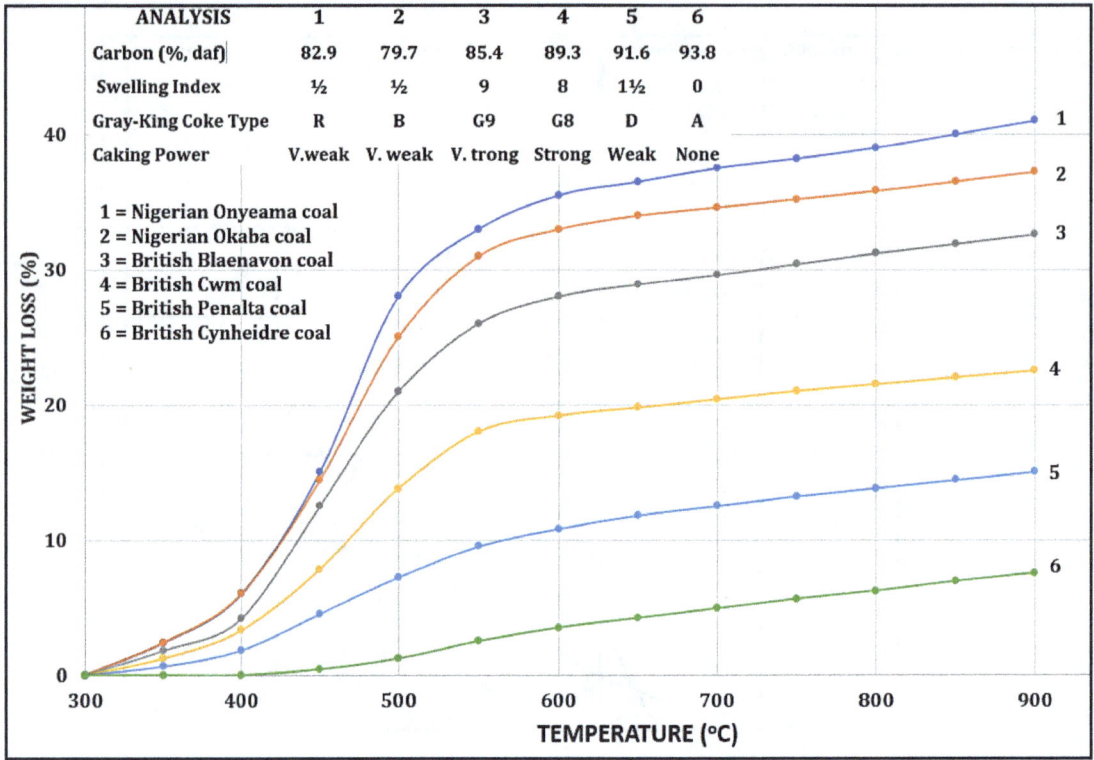

ANALYSIS	1	2	3	4	5	6
Carbon (%, daf)	82.9	79.7	85.4	89.3	91.6	93.8
Swelling Index	½	½	9	8	1½	0
Gray-King Coke Type	R	B	G9	G8	D	A
Caking Power	V.weak	V. weak	V. trong	Strong	Weak	None

1 = Nigerian Onyeama coal
2 = Nigerian Okaba coal
3 = British Blaenavon coal
4 = British Cwm coal
5 = British Penalta coal
6 = British Cynheidre coal

Figure 4.5 Thermograms of some Nigerian and foreign coals. (*Afonja,1974*).

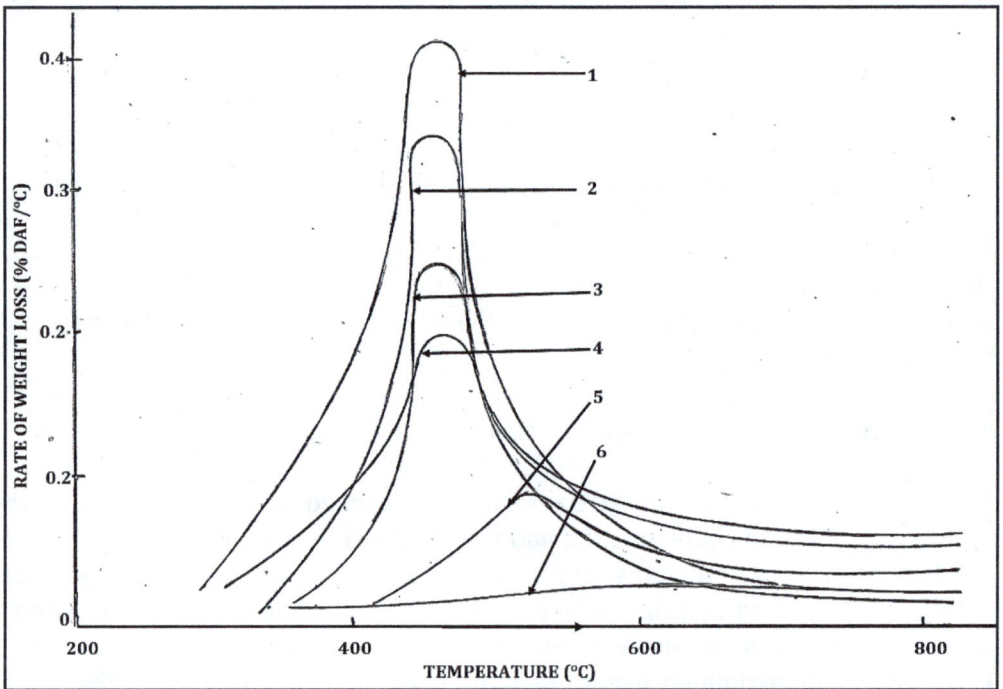

Figure 4.6 Rate of weight loss-temperature relationships for some Nigerian and foreign coals. (Sample numbers same as in Fig. 4.5) (*Afonja,1974*).

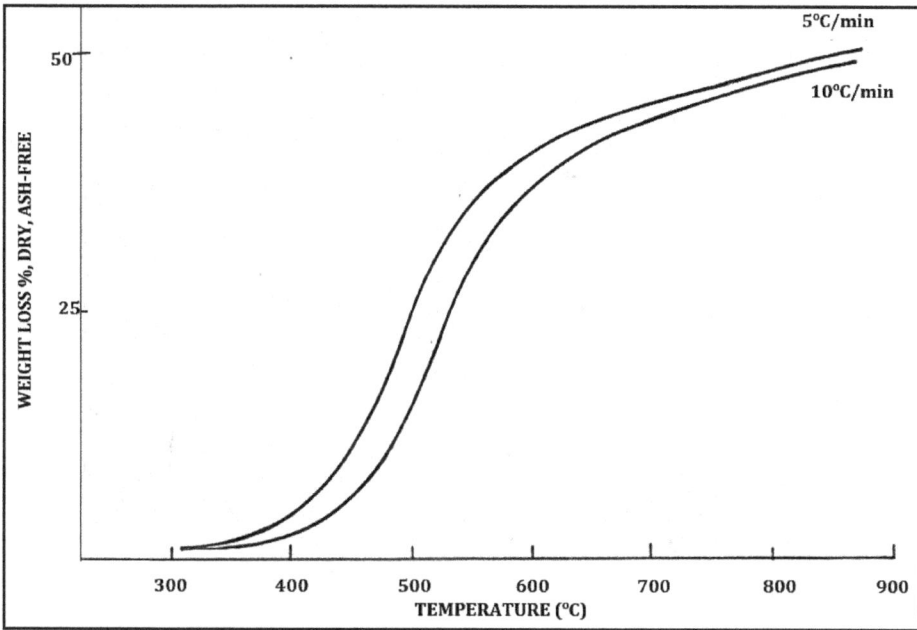

Figure 4.7 Thermograms for Obi/Lafia coal at two heating rates. *(Afonja,1979).*

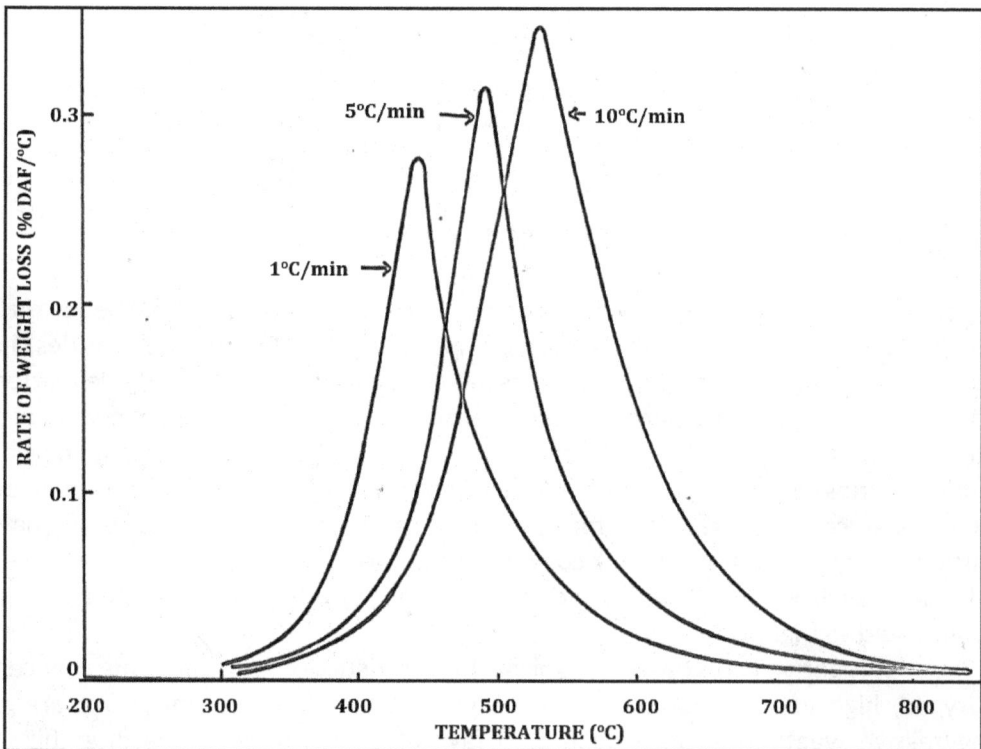

Figure 4.8   Effect of heating rate on the devolatilization rate of Obi/Lafia coal. *(Afonja,1979).*

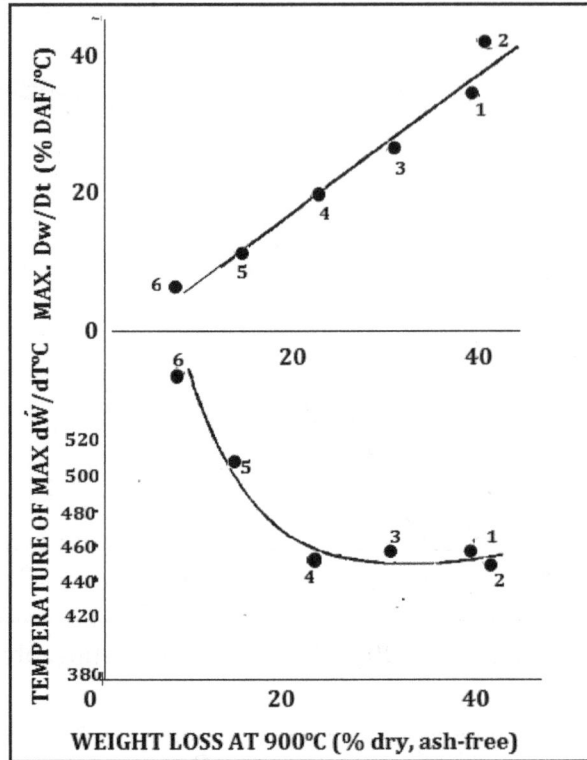

Figure 4.9   Rate of weight loss versus maximum weight loss for coals shown in Figure 4.5. *(Afonja,1979).*

4.4   WEATHERING OF NIGERIAN COALS

Weathering (oxidation) of coal occurs on exposure to the atmosphere. The rate is fastest for fresh coals and decreases with increasing oxidation. All the important properties change with oxidation, oxygen and moisture contents increase while carbon and volatile matter contents decrease. Calorific value of coal also decreases with oxidation and the coal becomes more friable. Weathering is believed to be caused by absorption of moisture and oxygen and the affinity varies with coal types and rank. In effect, an environment with generous supply of air and moisture is a prerequisite for coal oxidation. Low grade coals generally have greater affinity for oxygen than high rank coals and therefore tend to oxidize more when exposed to the same environment. Depending on the coal rank, type and environmental conditions, coal piles could self-ignite.

Most Nigerian coals have the typical characteristics of low rank coals: low carbon, high oxygen, high volatile matter, and high inherent moisture contents, all prerequisites for aggressive weathering on exposure to favorable environment. Despite this, there are significant variations between deposits in reactivity (with oxygen) and in the propensity for self ignition. Enugu coal is not known to be prone to self-ignition but Okaba coal is very susceptible. There were several fire incidents caused by spontaneous combustion involving Okaba and Orukpa coals in the late 1960s and early 1970s. Coal piles at the mines, in stockyards of consumers, and on ships during transportation to neighbouring countries self-ignited on many occasions. Furthermore, there were complaints by consumers that heating values did not meet delivery specifications. Other coal deposits were not being exploited hence not much was known about their self-ignition propensity. The fire incidents and consumer complaints prompted the Nigerian Coal Corporation to commission a study (Afonja, 1973). The investigation carried out at the Coal Research Laboratory (CRL) of University of Ife was designed to assess the proneness of Nigerian coals to weathering and self ignition, and also to determine the effect of weathering on the coal properties, in particular, the heating (calorific) value and slacking propensity. Apart from the local coals, eighteen foreign coals selected on the basis of their known reactivity to oxygen were also tested.

4.4.1   Assessment of proneness of Nigerian coals to self ignition

The crossing point method (Ganguly et al., 1953; Banerjee et al., (1972) was used to determine the susceptibility to self heating of four   Nigerian and eighteen foreign coals. Five-gram Samples (-100+75) were heated in a furnace within a reaction tube flushed first with nitrogen to remove entrapped air, followed by air stripped of moisture in a drying tower. A thermocouple was inserted just below the surface of the sample. The furnace was heated at a constant rate of 1°C/minute. The gas flow rate was kept low, at 80ml/minute to avoid disturbance of the sample. The temperatures of the sample and furnace were fed to a flat bed recorder. Heating was stopped when the sample temperature rose above the furnace temperature. The crossing point temperature (CPT) values are presented in Table 4.3. Low values of CPT indicate high propensity for self heating. However, several of the values did not agree with the known history of the coals. For example, Enugu coal considered to be the least reactive had the lowest crossing point (CPT) while Ogwashi lignite had the highest. The results obtained for the same coals by Ogunsola and Mikula (1991) were similar.

A second series of tests was designed to determine the glow temperature of the coals by heating samples in a modified Carbolite high temperature ash fusibility tube furnace fitted with a telescope for observation of samples during heating, a precision heating rate control system and a digital temperature output. The tests involved the heating of a coal sample in a shallow silica sample holder at a rate of 5°C/minute and moisture-free air at a flow rate of 80 ml/minute was passed through the furnace from a compressed air cylinder. A diffuser was installed at the entrance of air to the furnace to reduce the pressure and minimize dislocation of the sample. A thermocouple was immersed just below the sample surface and the digital signals from furnace and sample thermocouples were fed to a flat bed recorder. Calibration tests were run with two coals known to be prone and two known to be unreactive in order to establish the best temperature range for observation.

The temperature at which the sample glowed was taken as the ignition point. The point also corresponded to a sharp rise in temperature as recorded by the thermocouple immersed in the sample. In most of the tests on low grade coals, the point was also marked by an audible explosion. The results are also presented in Table 4.3 Low values of both crossing point and glow temperature indicate high reactivity. The results of the glow point tests were considered more reliable than the crossing point tests but there were still anomalies. For example, sample F5 known to be highly reactive had a higher glow point than F6 which is unreactive.

Table 4.3 Spontaneous ignition test of Nigerian and some foreign coals. *(Afonja,1973).*

SAMPLE	MOISTURE (%)	VM (%)	FIXED C (%)	O (%)	R_{omax} (%)	CPT (°C)	IGNITION (°C)	REACTIVITY HISTORY	FC/VM
				Low-grade coals					
N1-Ogwashi	37	59.2	40.8	21.3		163.5	345	unknown	0.69
N-2 Orukpa	6.0	44.0	56.0	12.9	0.53	156	366	prone	1.3
N-3 Okaba	5.0	43.2	56.8	11.8	0.48	158	355	prone	1.3
N-4 Enugu	2.1	42.6	57.4	5.9	0.62	160	420	highly reactive	1.4
F1	22	44.1	55.9		0.41	>200	356	prone	1.3
F2	37.5	38.1	23.1		0.32	>200	332	unreactive	0.6
F3	37.3	32.3		22.5		>200	356	unknown	0.9
F4	5.5	38.1	61.9	9.6	0.52	175	459	reactive	1.6
F5	3.5	41.1	58.9	10.6	0.54	158	516	highly reactive	1.4
F6	4.1	40.0	60.0	6.1	0.56	168	501	unreactive	1.5
F7	4.3	38.3	61.8	9.51	0.56	181	516	highly reactive	1.6
F8	3.5	38.7	61.3	9.71	0.57 ·	158	486	highly reactive	1.6
				High-grade coals					
F9	3.9	30.6	69.4	6.4	0.61	156	518	unreactive	2.3
F10	2.8	38.5	61.5	10.1	0.68	161	552	unreactive	1.6
F11	7.4	41.4	58.6	12.0	0.68	145	411	prone	1.4
F12	2.2	34.3	65.7	9.2	0.77	178	521	mildly reactive	1.9
F13	5.5	38.0	62.0	10.2	0.60	176	487	reactive	1.6
F14	5.1	30.7	69.3	10.0	0.67	184	544	unreactive	2.3
F15	3.5	32,7	67.3	8.7	0.96	178	527	unreactive	2.1
F16	0.9	24.8	75.2	4.3	1.04	184	541	unreactive	3.0
F17	1.0	11.0	89.0	1.5	1.64	190	481	unreactive	8.1
F18	1.6	4.8	93.6	0.3	3.1	>200	605	unreactive	19.5
				Coke					
F19		1.1	98.9				601		

A third set of tests was then carried out on a Linseis horizontal thermobalance (Afonja, 1973, 1974). Four foreign coals were tested along with the four Nigerian coals. The foreign coals selected are known to be prone, highly reactive or unreactive (Table 4.3). A sample of high temperature coke was also tested (F19). The weight -temperature profiles are shown in Figure 4.10.

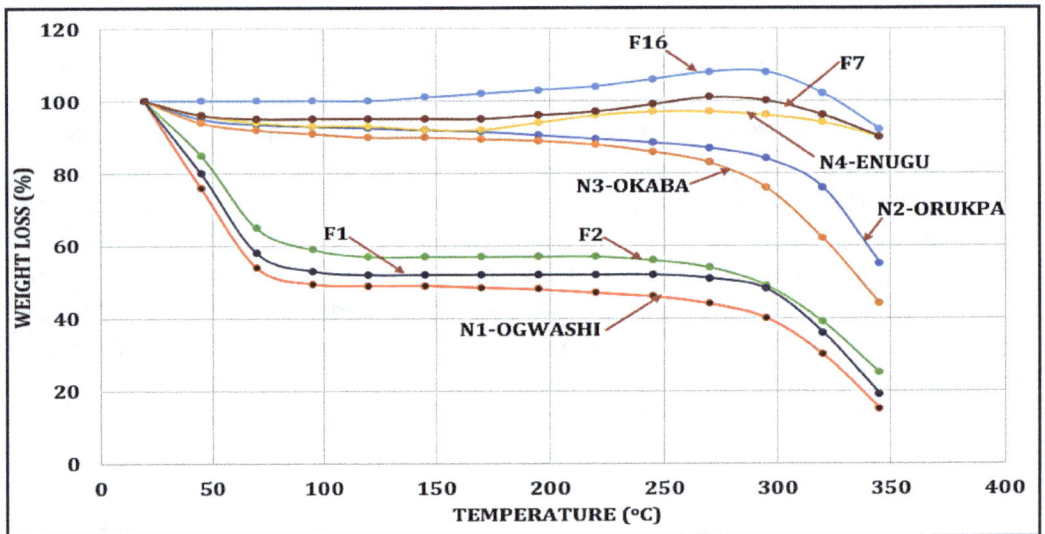

Figure 4.10 Thermogravimetric curves for Nigerian and foreign coals: Ogwashi lignite, Okaba sub-bituminous, Orukpa sub-bituminous, Enugu sub-bituminous, F1: foreign reactive lignite, F2: foreign unreactive lignite, F7: foreign reactive bituminous, F16: foreign unreactive low-volatile bituminous *(Afonja, 1973).*

Three of the coals known to be prone to self ignition (N2, N3 and F1) all had glow points less than 400°C. However, coal F7 known to be highly reactive had a value of 516°C while a lignite (F2) known to be unreactive had the lowest glow point of 332°C. The inability of both the crossing point and glow point tests to predict susceptibility to self ignition accurately is due apparently to the assumption that the point at which reactions in the coal become exothermic and temperature continues to increase without external heating source is also indicative of imminent self ignition.

The crossing point actually marks the beginning of sustainable self heating but actual ignition will depend on the types and kinetics of exothermic reactions taking place in the coal subsequently. Numerous investigations have shown that the reactions are quite complex and still largely not understood. However, it has been established that complexes formed at the lower temperature endothermic stage (which are still not well identified) and inorganic matter of the coal start to break down at the exothermic stage producing gases, mainly carbon dioxide, carbon monoxide, hydrogen, and hydrocarbons, mainly methane. All the gases with the exception of carbon dioxide are combustible and the relative amounts released would probably determine the ignition point of the coal. All the gases have higher ignition points than the residue coke but the glow of the coke could ignite the gases prematurely and start off combustion. The amount and chemical composition of gas released will depend on the coal rank and type. Low grade high volatile coals will release more gases and have low fuel ratios (fixed carbon:volatile matter ratios).

Low fuel ratio may be indicative of high coal reactivity. All the coals with ratios not greater than 1.6 in the test series were either prone or highly reactive, with the exception of coal F2 (Table 4.3). Recent work by Sahu et al. (2009) has also shown a strong correlation between volatile matter content and susceptibility to self ignition. While the anomaly of coal F2 could not be explained at the time of the investigation, the results of numerous investigations since then, particularly by Avila (2012) have provided feasible explanations. As a result of the very high moisture content of coal F2, evaporation takes place over a wide range of temperatures, resulting in weight loss and this tends to falsify the crossing point temperature (Figure 4.11). The evaporation also masks reactions due to adsorption of oxygen on the coal surface, the formation of coal-oxygen solid complexes and some weight gain. Furthermore, low-grade coals that already have high levels of oxygen content will have relatively low capability for further absorption.

Nadins is a low rank, high volatile bituminous coal known to be prone to spontaneous combustion while Goodehoop coal is a high rank, low volatile bituminous coal known to be unreactive. The moisture content of Nadins coal is 7.4% compared with 2.3% for Goedhoop coal. This explains the differences in the shapes of the thermograms for the two coals. Both the Nadins and Goedhoop coals show an inflection at around 100°C when moisture evaporation takes place. However, the Goedhoop coal temperature then rises steeply while moisture in Nadins coal continues to evaporate over a wider temperature range after which the heating curve rises more sharply than Goedehoop coal. In effect, the crossing point of Nadins coal is higher but the self heating rate is also higher once the moisture has been expelled.

The effect of high moisture may also explain the behavior of coal F2 which is known to be unreactive, based on the high crossing point temperature. However, the glow point is the lowest of all coals tested. The moisture desorption reactions appear to dominate and suppress coal reactivity. However, once the moisture has been expelled, self-heating rate becomes very rapid, hence the low glow point. Fuel ratio and crossing point temperature data for Nigerian coals are plotted in Figure 4.12, in addition to data for other coals drawn from CRL-OAU database. The four coals that are out of trend are low rank, high moisture, high oxygen coals which are now known to give false crossing point temperatures.

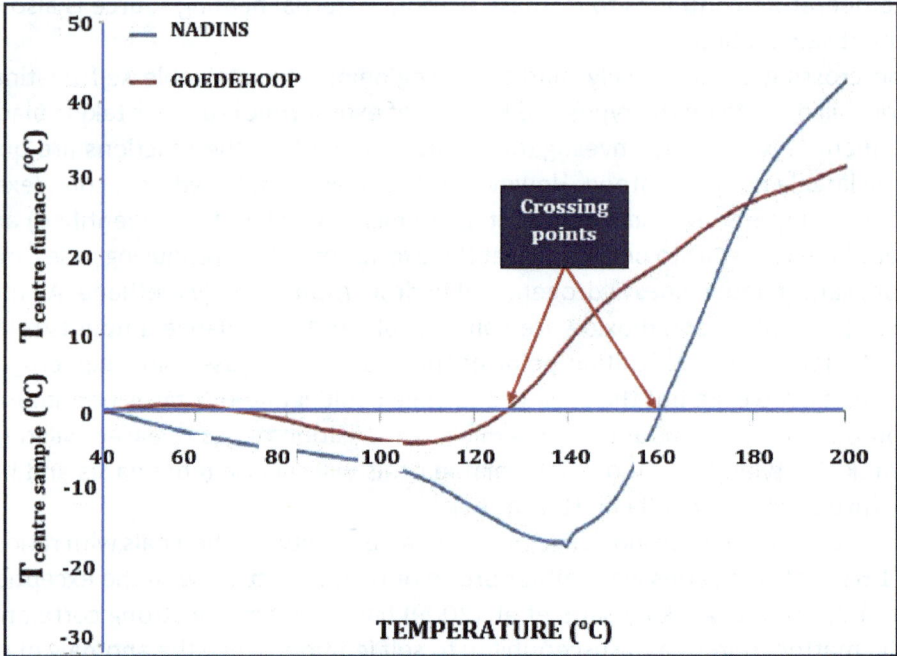

Figure 4.11 Differential thermal profiles for a low-rank and a high-rank coal.*(Avila, 2012).*

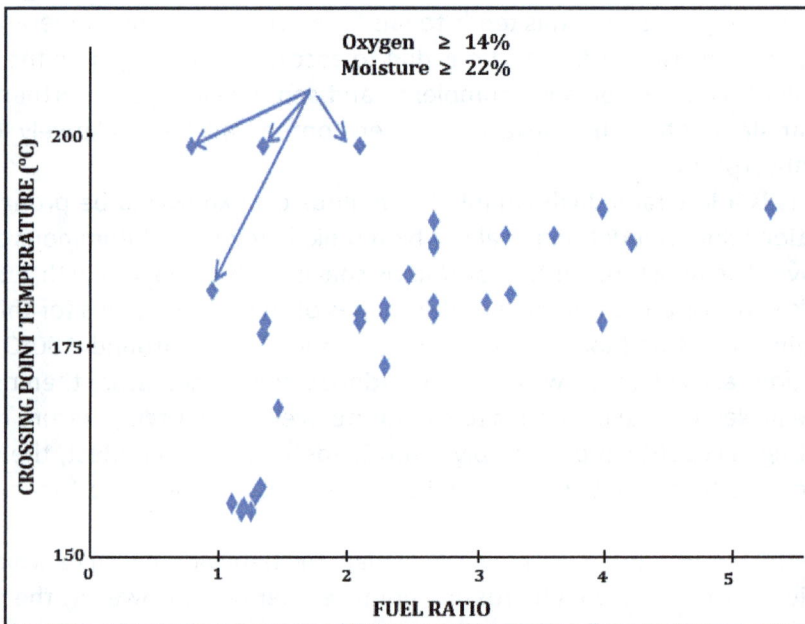

Figure 4.12 Relationship between fuel ratios and crossing point temperatures. *(Afonja, 1973).*

In a recent study Ahmad et al., (2009) also found that proneness to oxidation decreases with increase in fuel ratio. They showed that volatile matter is mostly the aliphatic and alicyclic part of coal while fixed carbon is the aromatic component. In effect, a high fuel ratio means more aromatic structure than aliphatic and alicyclic structures which enhance oxidation. It was found that coals containing high oxygen functional groups also had the lowest crossing point temperatures and therefore were highly prone to self-ignition. It has been established that coals of different ranks, types and origin have different capacities for absorption of oxygen. The results of numerous studies have shown also that high oxygen absorption capacity of coal is indicative of susceptibility to spontaneous combustion.

The chemical reactions involving oxygen in low temperature oxidation of coal (below 170°C) are still not well understood. However, there are several theories. Mazumdar et al. (1959) concluded that oxygen attacks the methylene groups constituting the hydro-aromatic structure of coal, forming peroxides and hydro-peroxides. Ahmad et al. (2009) found that high oxygen sorption by coal results in the formation of high proportion of oxygen functional groups, in particular, hydroxyl, carbonyl and carboxyl groups. Most of these reactions require the presence of adequate moisture. The formation of hydroxyl and carbonyl groups presents new sites for oxygen attack, resulting in the production of carboxyl and carbonyl groups.

In summary, Okaba and Orukpa coals are prone to self combustion, apparently because of their high propensity for oxygen and moisture absorption. Although Enugu coal is also reactive, the coal is less prone to spontaneous ignition because of the relatively low oxygen and moisture content. Ogwashi lignite is unreactive, based on crossing point temperature but prone to self-ignition on the basis of glow temperature, considering the very high moisture content and the discussion above on Nadins coal.

4.4.2   Assessment of the effect of weathering on slacking and heating value of Nigerian coals

In spite of numerous investigations dating back over a hundred years, the effect of weathering on coal heating value is still unclear. Tests carried out in Europe (King, 1905) indicated that coals could lose up to 30% of heating value in just six weeks of storage in open yards. Extensive tests on a wide range of coals of different ranks by Porter and Ovitz (1917) found a maximum deterioration of less than 6%, even for low grade coals stored barrels in open air for several years. La Grange (1950) suggested that the loss could be as high as 50% for low rank coals. The inference that can be drawn from the numerous published results is that deterioration of heating value is coal specific and depends on many variables, including the coal rank and type, the volume and chemistry of the volatile matter content, and the chemical reactions that take place during weathering. These reactions depend on the organic and inorganic constituents of coal as well as the storage conditions.

4.4.2.1   *Slacking of Nigerian coals due to weathering*

All coals have some tendency to absorb moisture and oxygen on exposure to favourable environment. However, propensity varies depending on the coal rank. Low grade coals have great affinity for moisture and tend to retain maximum inherent moisture relative to the environment. Changes in temperature will cause moisture to evaporate, leaving porous sites for the adsorption of oxygen. When ambient moisture increases, for example due to rain, the coal reabsorbs moisture. Changes in environmental temperature and moisture cause volume changes which vary from the surface to the core and thereby set up internal stress gradients which tend to break up the coal and expose more surfaces for oxygen adhesion and oxidation.

Tests were conducted on Enugu, Okaba and Orukpa coals to determine the response of the coals to weathering. Samples were taken from freshly mined coals, with size consist similar to consignments delivered to customers. The samples were stored in 25-kg barrels on open air for 52 weeks. The barrels had no lid and holes were drilled at the sides to allow air circulation from different directions as would occur in a typical stockyard pile. Samples were taken out every four weeks to determine calorific values and, at the end of the tests, screen analysis of the coals was carried out to determine the degree of slacking. All the coals decrepitated and degraded in size considerably over the storage period (Table 4.4). The minus 1 mm size increased by 300% for Enugu coal compared with nearly 2000% for Okaba coal.

Most low grade coals including Okaba and Orukpa coals have both high inherent moisture and oxygen content (Table 4.2) and are therefore prone to oxidation and degradation on exposure the favourable environment. Okaba coal is particularly susceptible to decrepitation and this explains why the coal was involved in most of the reported fire incidents. Apart from high moisture and oxygen, the coal also contains significant amounts of the maceral, ulminite in which extensive fractures are visible (Figures 3.3 & 3.5). Slacking causes coal to break up, exposing new surfaces to oxidation and leading to even more slacking and increased propensity for self-ignition.

4.4.2.2   *Loss of heating value of Nigerian coals due to weathering*

Enugu coal lost about 11% of heating value after 52 weeks of storage, 6% in the first 8 weeks. Okaba coal had lost about 25% by the end of the test period, around 20% in the first ten weeks of storage (Figure 4.13). Oxygen content increased and volatile matter, carbon and hydrogen contents decreased. All the changes were faster and more significant for Okaba coal. There was no significant difference in all the values for Orukpa and Okaba coals.

Table 4.4 Slacking of weathered Nigerian coals. *(Afonja, 1973).*

Size range (mm)	Enugu			Okaba			Orukpa		
	Initial (%)	Final (%)	% Change	Initial (%)	Final (%)	% Change	Initial (%)	Final (%)	% Change
20-16	20.5	16.4	-20.0	17.5	5.9	-66.3	18.7	8.3	-55.6
16-1	72.7	56.4	-22.4	78.9	20.3	-74.3	75.4	28.3	-62.5
>1	6.8	27.2	+300	3.6	73.8	+1950	5.9	63.4	+975
	100	100		100	100		100	100	

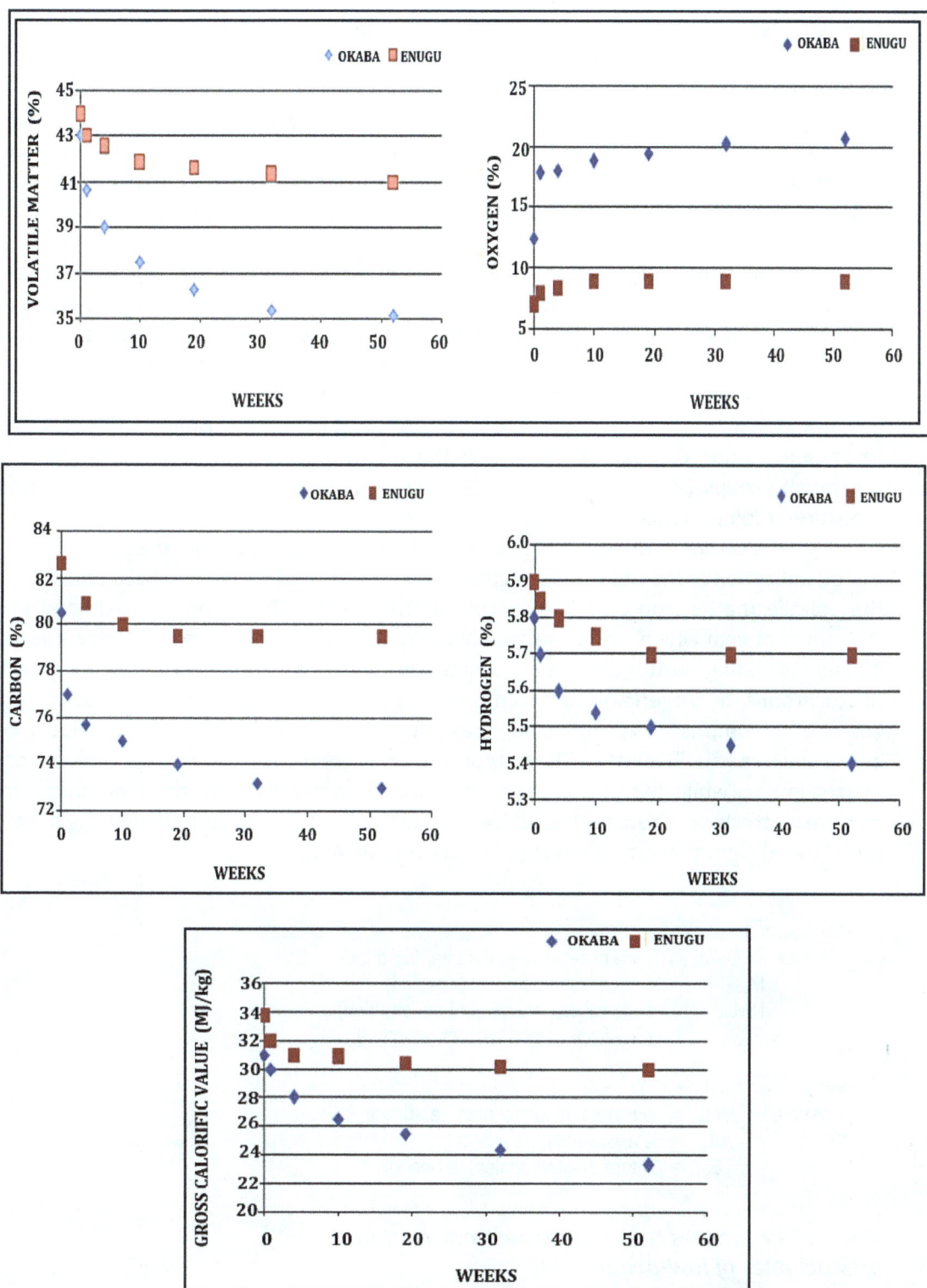

Figure 4.13 Effect of weathering on proximate/elemental analysis and heating values of two Nigerian coals. *(Afonja, 1973).*

All coals have a tendency to weather and disintegrate when alternatively wetted and dried in stockpiles by sunshine and rain. However, the effect on high-rank coals is insignificant. Bituminous coals lose 1-3.5% when exposed for a period one year or more but low-grade coals (lignite and sub-bituminous coals) can lose up to 20% of heating value in the first few weeks of exposure. The difference in weathering behavior between high-rank and low-rank coals is due primarily to differences in affinity for oxygen and moisture. Low-grade coals have much greater affinity of oxygen and moisture, both of which are primary requirements for coal oxidation.

Most coals contain aliphatic, aromatic and non-aromatic structures, the younger the coal, the lower the ratio of aromatic to non-aromatic structures. The non-aromatic structures in low-grade coals are predominantly aliphatic and alicyclic. The non-aromatic structures are particularly sensitive centres of attack of oxygen and oxidation will result in the formation of significant quantities of carbon monoxide and carbon dioxide (Fryer et al., 1973). There will be corresponding reduction in the volatile matter, carbon, and hydrogen contents as well as the heating value of coal. Furthermore, the increased concentration of oxygen-bearing functional groups (-OH, -COOH, etc) resulting from oxidation also tends to increase the moisture-holding capacity of oxidized coals.

Aromatic carbon is the primary source of the heat value of coal and the fixed carbon gives a good indication of the heat value of coals. However, part of the heat value also derives from the volatile matter content. The volatile matter derives from the aliphatic and alicyclic structures of coal which oxidize relatively easily and produce carbon monoxide and carbon dioxide. Volatile matter in coal is made up partly by combustible gases (methane, hydrocarbons, hydrogen and carbon monoxide), often referred to as active volatile matter. The balance comprises incombustible gases (nitrogen, carbon dioxide, sulphur oxides), known as the inactive volatile matter. The bulk of the active volatile matter derives from the organic matter in coal while the inactive component derives mainly from the inorganic,   mostly mineral matter) constituents. The active volatile matter contributes to heat value of coal and is estimated from the Parr formula (1917), (Equation 4.1).

$$OVM_{active} = 100 - (MM_{ad} + FC_{ad})$$ [4.1]

Where: OVM_{active} = % organic volatile matter (air-dried)
   M_{ad} = % moisture content (air-dried)
   MM_{ad} = % mineral matter content (air-dried)
   FC_{ad} = % fixed carbon content (air-dried)

$$MM_{ad} = 1.08\,A_{ad} + 0.55\,TS_{ad}$$ [4.2]

Where MM_{ad} = % mineral matter content (air-dried)
   A_{ad} = % ash content (air-dried)
   TS_{ad} = % total sulphur content (air-dried)

4.4.2.3   *Proneness to spontaneous combustion of stockpiles of low-grade coals*

It has been established through many investigations that when coal is oxidized at a low temperature, the adsorption of oxygen leads to the production of peroxides first, and the peroxides decompose into carbon monoxide and carbon dioxide and water. It is known that, as the integrated amount of oxygen adsorbed on the coal increases, the number of adsorption sites decreases, lowering the rate of adsorption. This means that, in a pile, the heat generating reaction proceeds by consuming oxygen in the air and, as time passes, the activity of this reaction decreases, slowing down the heat generation. Low grade coals have greater affinity for oxygen and the oxygen consumption rate (OCR) is higher than for bituminous and other higher-rank coals. The OCR increases exponentially with increasing temperature but the reaction rate decreases for all ranks of coals as the integrated value of oxygen increases (Figure 4.14).

Moisture is present in coal in two forms: moisture adsorbed in the pores of the coal and moisture from external sources such as rainfall which adheres to the surface of the coal. Adhered moisture evaporates much faster than adsorbed moisture, at a rate controlled by the relative humidity, pressure and the amount of the adsorbed moisture. The moisture contained in a coal pile evaporates, causing the latent heat of evaporation to reduce or cancel out the heat generated by low-temperature oxidation, thus lowering the temperature inside the pile. The temperature in the coal pile depends on the net effect of the three competing reactions: adsorption, desorption and evaporation. Again low-rank coals absorb more moisture than higher grade coals for a given relative pressure (Figure 4.15). The high adsorption and desorption rates will induce high thermal stresses in low-rank coals, causing decrepitation and exposure of new surfaces to oxidation reactions.

Okaba coal has the greatest affinity for moisture and oxygen of all the Nigerian low rank coals and is also the most fryable of all the coals tested. Intermittent absorption and desorption of moisture will decrepitate the coal and expose more new surfaces to oxidation. If the prevailing environmental conditions do not favour fast removal of the net heat generated by oxidation, local temperatures in piles of the coal can rise rapidly and reach thermal runaway, resulting in self-ignition. It should be noted however that the dynamics of oxidation and weathering of coals in stockyards would be quite different from small-scale experimental coal piles. Coal is stored in stockyards in relatively large piles several metres high, holding 20 to 50 tonnes of coal. Much of the weathering takes place at and near the pile surface where air and moisture penetration is high.

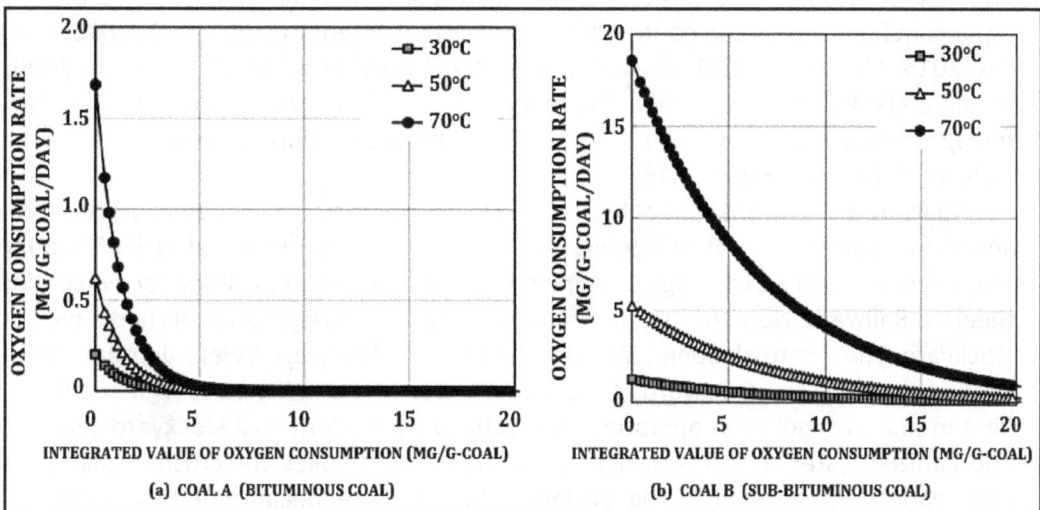

Figure 4.14 Oxygen consumption rates for different ranks of coal. *(Pak et al., 2015).*

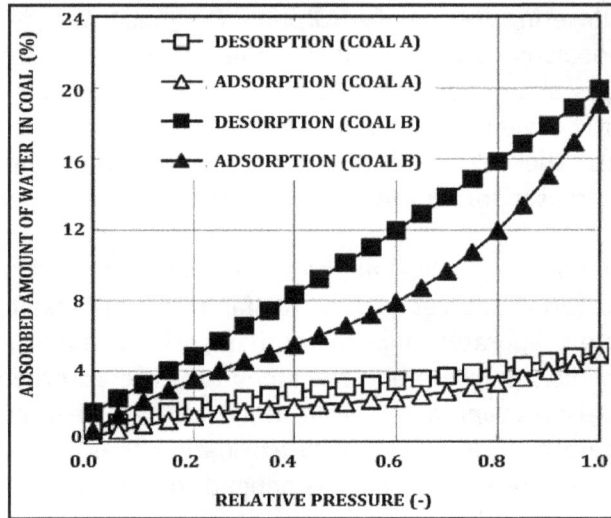

Figure 4.15 Water adsorption/desorption isotherms at 40°C for different ranks of coal. Coal A = Bituminous coal, Coal B = Sub-bituminous coal. *(Pak et al., 2015).*

While the overall heat build-up due to oxidation may be slow, the layers close to the surface of coal piles gain and lose moisture and heat generated due to oxidation relatively easily but the heat generated in the inner layers is often trapped and can build up rapidly. The high heat build up in the deeper layers of coal piles will enhance oxidative reactions and faster release of combustible gases which will result in faster loss of heating value. Also, temperature can rise to the point at which self ignition can occur in areas remote from the surface, even when at sub-critical near-surface temperatures. The design of a coal stockyard and the coal piles can have very profound influence on oxidation and propensity for self-ignition and a lot of research has shown that yard layout, ventilation, control of exposure to sunshine and rain, drainage control, ease of accessibility to coal piles, and the stockpile geometry all have significant influence in the control of proneness to self-ignition (Pak et al., 2015; Sloss, 2015; IEA, 2015; Cliff et al., 2014; de Korte, 2014; Chandralal et al., 2014; Coaltech, 2009; Okten, et. al., 2009; Fierro et al., 1999). The management of low-grade coal stockpiles during land transportation and shipment is also vital for the prevention of spontaneous combustion (Falcon, 2014). (See Figure 4.16).

Apart from the proneness of Nigerian coals to oxidation, the handling of coal at the mines and storage conditions at the points of utilization greatly enhanced loss of heating value. Freshly mined coal was stockpiled at mine sites for many weeks waiting for evacuation by the Nigerian Railways. None of the four major consumers of Nigerian coals had a well-designed stockyard. They were all open (uncovered) spaces and deliveries were dumped haphazardly. Coal stockpiles at the stockyards of consumers were around 2 metres high and there was no 'first in first out' policy in operation as practiced by modern coal stockyards. In view of the unorganized system of piling, forklift trucks trampled on piles and crushed coal, causing size degradation which enhances coal oxidation. Furthermore, because of the unreliability of coal delivery on schedule, consumers held large stocks for many months and this greatly enhanced the potential for loss of heating value and self-ignition.

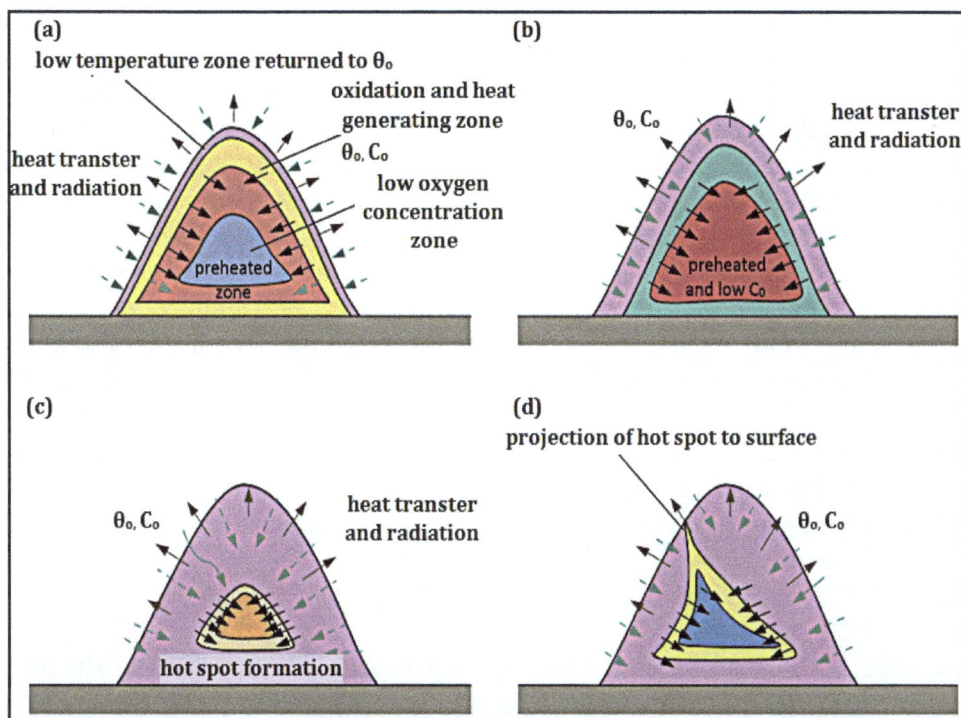

Figure 4.16   Example of how self-heating and spontaneous combustion can occur in a stockpile of low-grade coal. *(Sasaki et al., 2014).*

REFERENCES

Afonja, A. A. (1972). "An assessment of the coking quality of Nigerian coal seams." ENE/B/W/3. Report No. IFE/CHE/CP/1. For the Nigerian Steel Development Authority.

Afonja, (1973). "An investigation of the spontaneous combustion propensity of Nigerian Coals." Research Report: IFE/CRL/NCC/7. Nigerian Coal Corporation.

Afonja, A. A. (1974). Evaluation of the coking characteristics of some Nigerian coals by thermogravimetry. West African Journal of Science, 19(2), 171-74.

Afonja, A. A. (1979). "Further Studies on the Obi/Lafia Coal Deposit (Phase II)." Research Report No. IFE/CHE/CP/11, Coal Research Laboratory, Ife.

Afonja A. A. (2017). *Basic Coal Science and Technology.* SineliBooks.

Ahmad, I., Nabiullah, M. and R. S. Prasad (2009). " Role of oxygen functional groups and fuel ration in self heating of coal." *Journal of Mines and Fuels*, 57(1): 23-27.

Avila C. (2012). *Predicting self-oxidation of coals and coal/biomass blends using thermal and optical methods*. Ph.D thesis, University of Nottingham.

Banerjee, S. C., Nandy, D. K., Banerjee, D. K., and R. N. Chakravorty (1972). " Classification of coal with respect to their susceptibility to spontaneous combustion." *Transactions of the Mining and Metallurgical Institute of Indi.,* *V.* 59, No.2, 15-31.

Chandralal, N., Mahapatra, D., Shome, D. and P. Dasgupta (2014). "Behaviour of low rand high moisture coal in large stockpile under ambient condition." *American International Journal of Research in Formal, Applied and Natural Science*, 14-2010.

Cliff, D., Brady, D., and M. Watkinson (2014). "Developments in the Management of Spontaneous Combustion in Australian Underground Coal Mines." 14th Coal Operators' Conference, The Austr. Inst. Mining &Met. & Mines Assoc. Australia, 2014, Pp. 330-338.

Coaltech (2009). Prevention and control of spontaneous combustion. Coaltech Research Association. miningandblasting.files.wordpress.com/2009/09/spontaneous_combustion

de Korte, J. (2014). "Managing spontaneous combustion of coal.." CSIR. www.fossilfuel.co.za/confrence2014.

Falcon, R. (2014). "Self heating and spontaneous combustion in coal shipments." www.fossilfuel.co.za.

Fierro, V. et al., (1999). "Prevention of spontaneous combustion in coal stockpiles: Experimental results in coal storage yard." Fuel Processing Technology, Vol. 59, Issue 1, 1999, Pp. 23-34.

Fryer, J. F. and A. J. Szladow (1973). "Storage of Coal Samples." Info. Series NO. 66, Alberta Research Council, 1 (1973).

Ganguly, M. L., and N. G. Banerjee (1953). "Critical Oxidation and Ignition Temperature of Coal." *The Imma Review,* Vol. 2, No.4.

King, H. R. (1905). "Weathering of coal." *University of Illinois Bulletin*, Vol. IV, No. 33, August, 26, 1907.

La Grange, C. C. (1950). "Spontaneous heating of coal – summarized extracts from the literature." *Report No. 9 of 1950, Fuel Research Institute of South Africa.*

Lowry, H. H. (Ed.) (1963). Chemistry of Coal Utilization, Supplementary Volume. John Wiley.

Mazumdar, B. K. et al (1959). "Mechanism of the oxidation of coal." *Fuel*, Vol. 38, 469-482.

Ogunsola, O. I. and R. J. Mikula (1991). "A study of spontaneous combustion characteristics of Nigerian coals. " *Fuel* 70(2):258-261.

Okten, G. O. Kural and E. Algutkaplan (2009). "Storage of Coal: Problems and Precautions." In Energy Storage Systems – Vol. II pp. 172-186. Ed. Yalcin Gogus. UNESCO-EOLSS.

Pak, H. et al., (2015). "Evaluation of spontaneous combustion in stockpile of sub-bituminous coal." *211 KOBELCO Technology Reviews*, No. 33, Feb., 2015.

Parr, S. and W. Wheeler (1908). "The deterioration of coal." *Journal of American Chemical Society*, Vol., 30, pp. 1027-1033.

Parr, S. and F. Kressman (1911). "The spontaneous combustion of coal." *Journal of Industrial Engineering Chemistry,* pp. 152-158.

Parr, S. W. (1917). " Effects of storage upon the properties of coal. Bulletin No. 97, University of Illinois, Engineering Experiment Station.

Porter, H. C. and F. K. Ovitz (1917). "Deterioration in the heating value of coal during storage." Bureau of Mines, Bulletin 136.

Sahu. H. B., Mahapatra, S. S., and D. C. Panigrahi (2009). An empirical approach for classification of coal seams with respect to the spontaneous heating susceptibility if Indian coals." *International Journal of Coal Geology*, Vol. 80, pp. 175-180.

Sasaki, K et al. (2014) "Numerical modeling of low rank coal for spontaneous combustion." Proc., *Coal Operators Conference 2014*, Univ. Of Wollongong, South Africa.

Sloss, L. L. (2015). *Assessing and managing spontaneous combustion of coal.* CCC/259. iea.org.

Speight, J. G. (2012). *The Chemistry and Technology of Coal*, Third Edition. CRC Press.

van Krevelen, D. W (1961*). Coal*. New York, Elsevier.

5 Mining and beneficiation of Nigerian coals

5.1 INTRODUCTION

The value and utility of coal depend on many factors including chemical composition, heating value, ash content, response to cleaning processes, caking value, and weathering characteristics. The geology of the coal deposit, in particular, the seam depth, seam thickness and continuity, nature and geology of the overburden and seam intrusions are also vital variables which determine the technical and economic feasibility of mining the coal.

5.2 MINEABILITY OF NIGERIAN COALS

Coal seams are usually sandwiched between thick layers of mainly inorganic matter, shale, rock, clay etc. If the seam is shallow coal can be recovered by opencast mining which involves the removal of the overburden to expose the coal bed. Coal mined by this technique is relatively clean and often does not require further cleaning. Such coal usually does not have any caking properties and is primarily used for power generation and steam raising. Opencast mining technology is relatively simple and the most economic method of coal recovery. However, vast areas of land are effected and, unless there are laws which require reclamation of the land when the mine is closed, the negative impact on the environment can be quite severe.

Many coal seams are located several hundred metres below the surface and deep shaft mining is the only feasible technology, although a new technology which involves gasifying the coal in-situ without mining is already being deployed commercially. Deep shaft mining involves sinking shaft to the depth of the seam and tunneling to various faces. Miners use hydraulic or pneumatic hammers to break up coal which is loaded on conveyors and transported to the surface through shafts. Modern mines are mechanized and automated machines tunnel through coal seams to break up the coal. Loading of the coal onto conveyors and transportation to the surface are also automated.

5.2.1 Enugu coal mines

Coal deposits in the Enugu area are at substantial depths and there is a steeply rising scarp overlying the coal horizons. The coal was mined by one of two methods depending on the geology of the area, seam thicknesses and the thickness and nature of the overburden. In areas where the coal seam was over 1.5 metre and the overburden was in excess of 450 metres, the Room and Pillar (Pillar and Stall) system was adopted (Figures 1.21 & 5.1). The Room and Pillar involved creating a series of "rooms" connected by a network of pathways, using coal pillars as dividers and supports. The coal face was worked with pneumatic and hydraulic hammers, and evacuation of the coal to the surface was done by a rope haulage system. When the seam surface was exhausted, some of the coal pillars were blasted with explosives as miners retreated, to produce more coal. Wooden pillars and steel props were used to support the roof and sides during this process (Ugwu, 1996). The Room and Pillar mining method is crude, labour intensive and very dangerous. There were many accidents including roof collapse, flooding and explosions. In spite of this, it is the most appropriate technology in certain geological conditions.

Some coal seams at depths of 100 metres or more had long coal faces of 100-250 metres. The long wall process was adopted for winning the coal in such areas. Long tunnels were cut into the coal seam, using powered roof and side support systems. Cutting machines then worked the coal faces and the coal produced was evacuated to the surface by a system of conveyor belts. This method of mining which may be advancing or retreating is safer and more productive, and the yield is higher than in the pillar and wall system because no coal pillars are left behind.

Enugu coal deposit was mined for the first time in 1916, the same year in which the first rail line linking Enugu and Port Harcourt was constructed to facilitate the export. Production grew steadily, then collapsed due to the global economic depression of the 1930s. The second world was from 1939-45 and the need to fuel British warships and power stations provided new outlets for Nigerian coal.

Figure 5.1 Underground coal mining in Nigeria by the Room-and-Pillar method.
(Nigerian Coal Corporation).

In the meantime, internal demand grew and by the late 1950s the mine was supplying coal to the Nigerian Railway Corporation, Electricity Corporation of Nigeria and Nkalagu Cement Company. Coal was also exported to some West African Countries including Ghana and Egypt. Enugu coal deposit lies below a steeply rising scarp, hence the only mining technology option was underground mining. The coal was mined mainly at two faces - Onyeama and Okpara using a system of adits (straight roadways) and drifts (inclined roadways). However several other faces were opened at one time or the other to meet increasing demand. Mining was underground and both the Room-and-Pillar and Longwall techniques were adopted.

In the Third National Plan (1975-80) an attempt was made to mechanize the Enugu coal mines as part of the drive to promote more local consumption and develop an export market for Nigerian coals. There were many responses to the international call for bids but government decided to award the contract to Kopex, a Polish company which was not one of the bidders, as part of a Nigeria-Poland Cooperation Agreement. The Nigeria-Polish contract required that the mining process, coal evacuation system and coal storage facilities be fully mechanized. Other aspects of the contract included the installation of a coal washery. The project was a disastrous failure for many reasons, the main one being the incompetence of the contractor. The design of the mine was faulty and the quality of equipment supplied was poor. The post-windup coal production actually fell because new coal faces had to be developed and manual systems of winning re-adopted (see Ugwu, 1996).

5.2.2   Okaba and Orukpa coal mines

Okaba and Orukpa deposits are located north of Enugu and both were mined for many years. The overburden of both deposits are relatively thin, less than 50 metres in some areas. The topography of both areas and the coal-to-rock ratios of between 1:7 and 1:1 favoured the opencast mining technique. Both deposits were exploited by the opencast method which involved the removal of the overburden to reach the coal seam. Mining of Orukpa coal stopped with the decline of the demand for coal but the Okaba mine has remained active, operating on and off.

5.2.3   Other coal mines

Apart from the mines in Enugu and Okaba zones, no other coal mine has been developed, although several outcrops in Ogwashi-Asagba, Gombe and other areas have been exploited locally by opencast mining, mainly for the production of cement. However, with the recent privatization of ownership of most of the deposits, activities are in progress to develop several of the mines.

Obi/Lafia coal deposit has not been fully investigated and the data currently available on the extent of the deposit and technical properties derived from borehole drilling in the early 1970s. All the seams identified were too deep for opencast mining. Most of the seams identified were considered too thin or sporadic for underground mining. Only seams 12 and 13 of the deposit were considered sufficiently thick and most of the analysis done so far have been on samples from these two seams. The geology of the area and the proven mineable reserves are not considered favorable for mechanized mining. Furthermore, the sulphur content is excessively high and the response to the coal to cleaning is poor. The Adamawa deposit has similar characteristics but there has be no systematic evaluation of the resource.

5.3   HANDLING CHARACTERISTICS
OF NIGERIAN COALS

Mechanized coal mining always produces considerable gangue along with coal, hence cleaning is often necessary. If the coal contains a significant amount of sulphur of the order of 1% and above, the level must be reduced to upgrade the value of the coal in view of increasingly stringent restrictions on the release of sulphur oxides into the environment. Also, international trade in coal is substantial and increasing rapidly, and intercontinental transportation is now quite common. Furthermore, coal for use in power generation and blast furnace injection is pulverized to enhance handling, transportation and feed automation.

Resistance to weathering is also vital if the coal is to be stored for any length of time during transportation and prior to use. The economic value of a coal is therefore determined by ease of mining, resistance to degradation on handling, washing and grinding characteristics, resistance to weathering, and the heat value per unit weight or weight of coal required per unit energy generation. The ability of the coal to withstand washing processes, handling and transportation is vital therefore in assessing its value. Three parameters are important: handling and transportation stability, grindability, and washability.

5.3.1   Handling and transportation characteristics
of Nigerian coals

Extensive studies have been carried out on Nigerian coals to determine the grindability and washability of Nigerian coals (Estel, 1996, Afonja, 1977a, 1977b, 1979, Roe, 1982), although not all the coals deposits were tested and not all the parameters were investigated. Available data shows that Enugu and Orukpa coals have very good resistance to handling degradation. Okaba coal is also resistant initially but tends to weaken on exposure due to oxidation leading to poor handling characteristics. Storage for extended periods can cause lump coal to lose vital properties, in particular the heating value, and to disintegrate completely into fine powder. One of the major complaints of the main customers of the defunct Nigerian Coal Corporation was the rapid deterioration of the heating value of Enugu and Okaba coals in storage. If the environmental conditions are favourable (inherent coal properties humidity pile height, ventilation, etc), coal can ignite spontaneously in storage. There were indeed several instances of spontaneous ignition of Okaba coal during storage as well as during transportation to a West African country (See Chapter 4).

5.3.2   Grindability of Nigerian Coals

All the planned coal-fired power generation plants in Nigeria will include a fluidized-bed combustor, hence the coal must be pulverized. The grindability of some Nigerian coals has been discussed in Chapter 3. The coals from Enugu zone are hard, strong and do not decrepitate appreciably on handling. The coals are harder than typical sub-bituminous coals and specific grinding energy may be high. Okaba coal is relatively soft and the size consist degrades both with handling and exposure to oxidative environment. It should be relatively easy to pulverize this coal for use at the power plant under construction near the deposit. The is no available data on the grindability or handling characteristics of the other known coal deposits.

5.4   CLEANING CHARACTERISTICS OF NIGERIAN COALS

Coal is washed to reduce associated mineral matter, in particular, ash-forming constituents and sulphur. The coal must be ground to a specific size range depending on the method of cleaning, and dewatered after cleaning. In some cases, further air drying may be required. The most common method is the sink and float process based on the difference between the specific gravity of coal and the associated gangue. Pure bituminous coal has a specific gravity of about 1.45 compared with 4.5 for shale. Washing the coal in fluids of suitable specific gravity will cause the heavier material to sink while the relatively clean coal floats. The proportion that sinks contains free ash distributed in discrete bands in the coal while mineral matter from the original coal-forming plant material is retained in the floats as inherent ash.

5.4.1   Washability characteristics of Nigerian coals

The sulphur and ash contents of Nigerian coals are shown in Table 5.1. None of the Nigerian coal deposits that had been mined commercially had a washery. Because mining was manual, the ash level of coal produced was low and well within acceptable limits for power and heat generation. Three of the four mines (Okaba, Orukpa, Ogboyoga) being developed currently under private ownership are opencast mines, hence cleaning may not be required. Ezimo coal deposit can be mined by either opencast or underground technique. However, all the deposits in the Enugu zone are deep and can only be recovered by underground mining. It is not clear however whether mining will be mechanized. If so, the coal will need to be washed to meet current international standards for ash content, especially if some of the coal produced will be exported. The sulphur levels in all the coals with the exception of Obi/Lafia coal are also very low and, for most applications processing specifically to reduce sulphur content would not be required.

     Washability curves for Nigerian coals are shown in Figures 5.2-5.8 (Afonja, 1976, 1977a, 1977b). Enugu coals (Onyeama, Okpara, and Ribadu) show good cleaning characteristics, with about 70% low-ash (about 7%), clean coal recovery at specific gravity of 1.3. Okaba coal yield at same specific gravity is similar but the ash content is slightly higher at about 8%. Orukpa coal has the best cleaning characteristics, with about 90% yield of coal of about the same ash content at the same specific gravity. The very steep cumulative float graphs for the coals from the Enugu-Okaba belt indicate a low spread in ash content and low sensitivity to specific gravity. In effect, increase in ash removal with increase in specific gravity of the cleaning fluid will be minimal, an indication that much of the ash is intimately embedded in the coal mass and only the discrete lumps are removed by washing. The shapes of the fractional float curves for Enugu-Okaba belt coals show that the coals can be easily cleaned with good yield at relatively low specific gravities. The cumulative float curves show that a

Table 5.1 Ash and sulphur levels in Nigerian coals. *(Arthur D.Little, 1965; Afonja, 1972, 1975, 1976, 1979).*

QUALITY PARAMETER (%, dry)	ENUGU	INYI	ORUKPA	OKABA	LAFIA (WASHED)	OGWASHI
Ash	3.5-21.6	6.5	8-9	9.6-15.9	14.7	14.4
Total Sulphur	0.75	1.47	0.46	0.7	1.5-6.5	0.78
Sulphate Sulphur	0.01	0.02	0.02	-	0.015	0.05
Pyritic Sulphur	0.05	0.75	0.02	-	1.4-6.5	0.02
Organic Sulphur	0.69	0.67	0.43	-	0.07-0.3	0.71

high percentage, over 70% of clean coal can be recovered at specific gravity below 1.5, with the exception of Ribadu coal which would require higher specific gravity for the same level of yield. The curves for Obi/Lafia coal showed a wider ash variation with increase in recovery, with the fractional and cumulative float curves approximating to straight lines. This is indicative of poor response to cleaning.

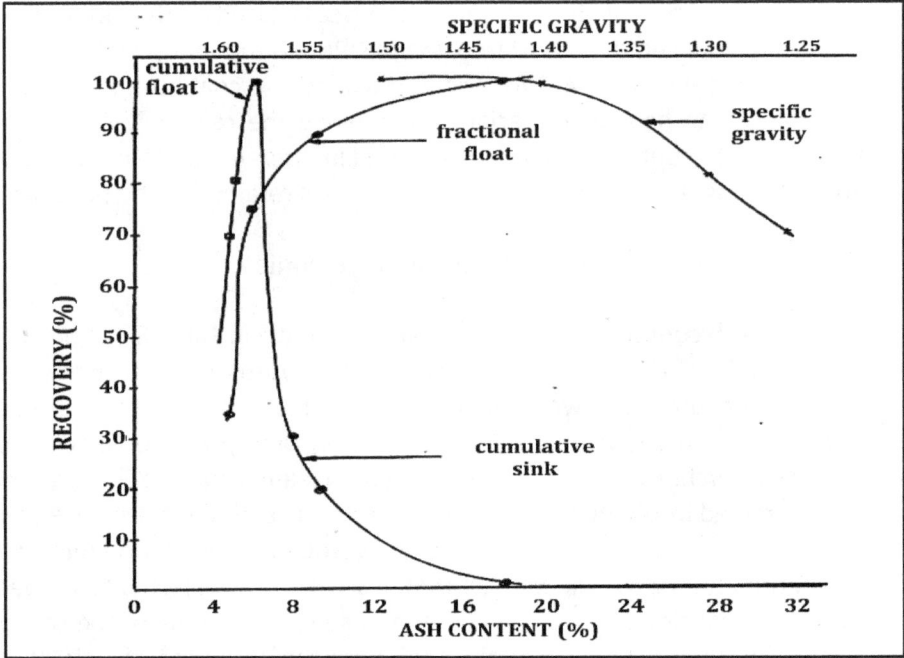

Figure 5.2 Washability curves for Onyeama coal. *(Afonja, 1976, 1977a&b).*

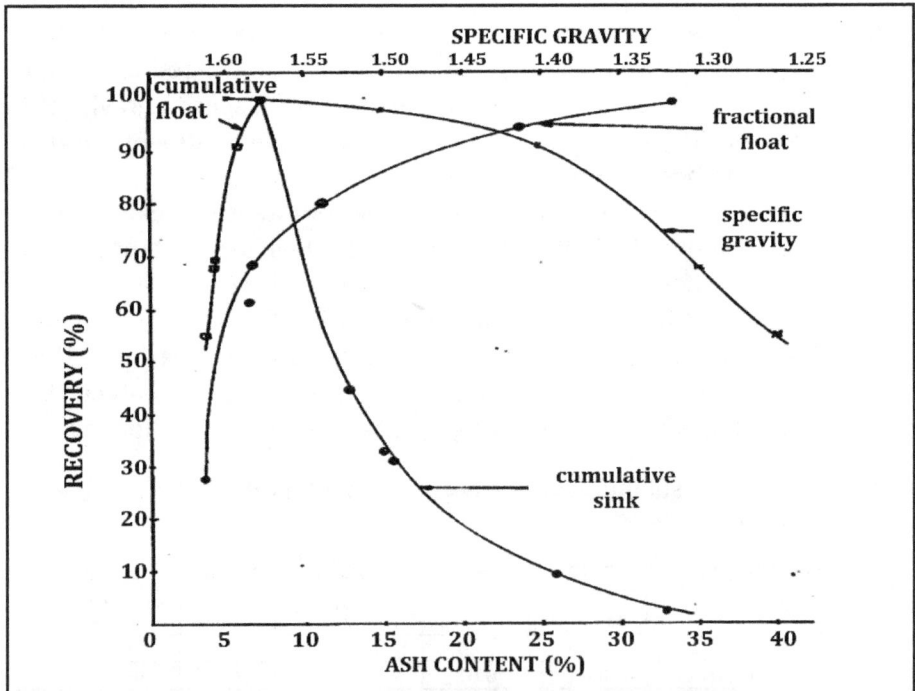

Figure 5.3 Washability curves for Okpara coal. *(Afonja, 1976, 1977a&b).*

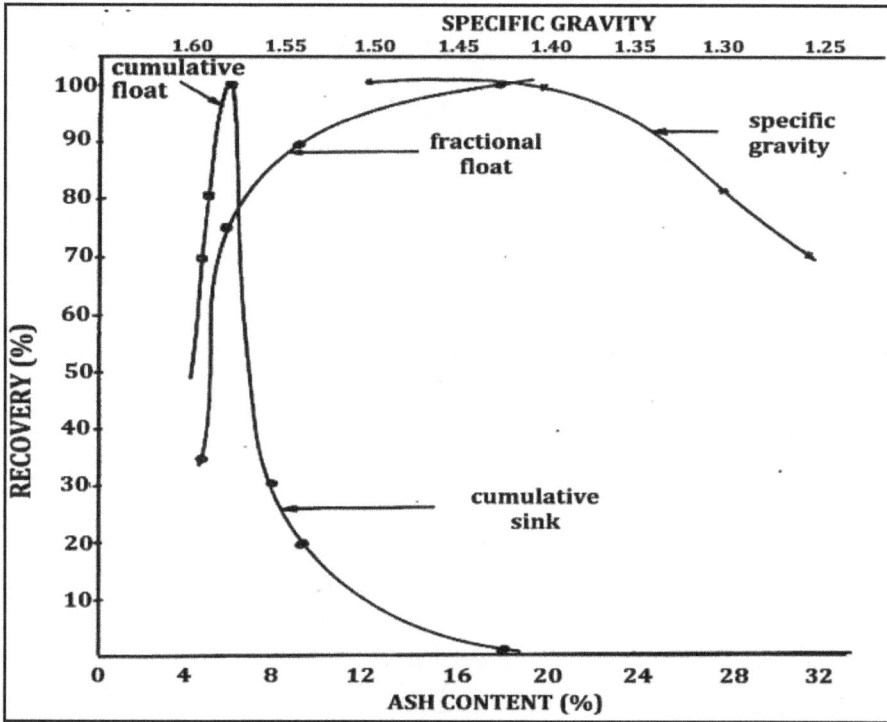

Figure 5.4 Washability curves for Ribadu coal. *(Afonja, 1976, 1977a&b).*

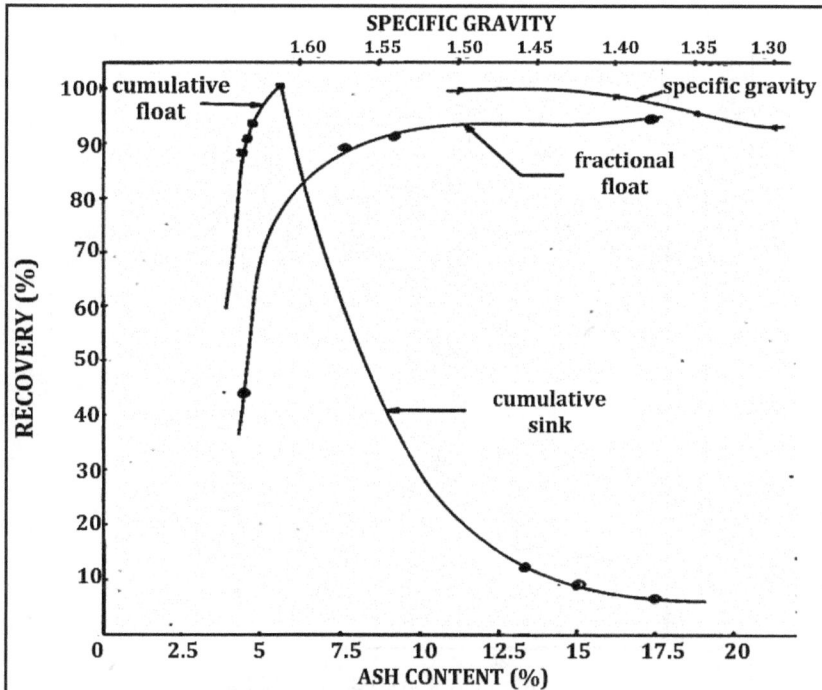

Figure 5.5 Washability curves for Orukpa coal. *(Afonja, 1976, 1977a&b).*

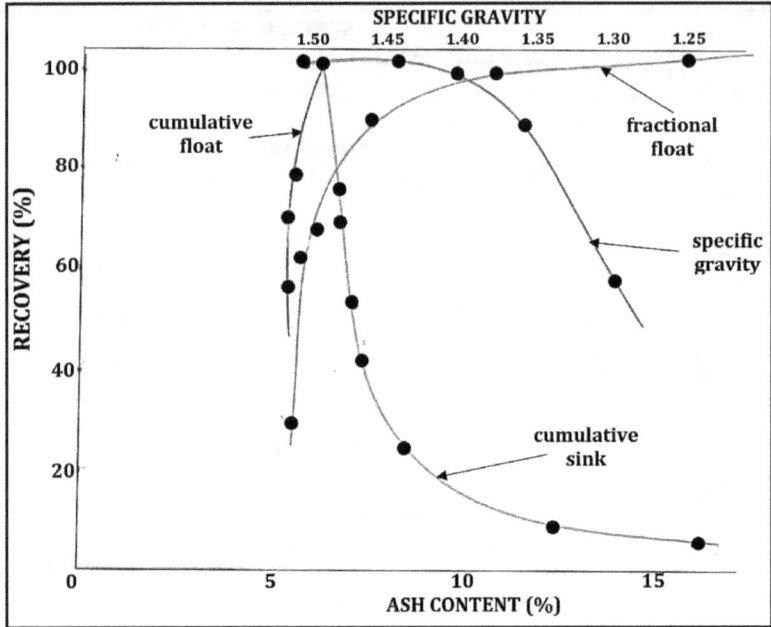

Figure 5.6 Washability curves for Okaba coal. *(Afonja, 1976, 1977a&b).*

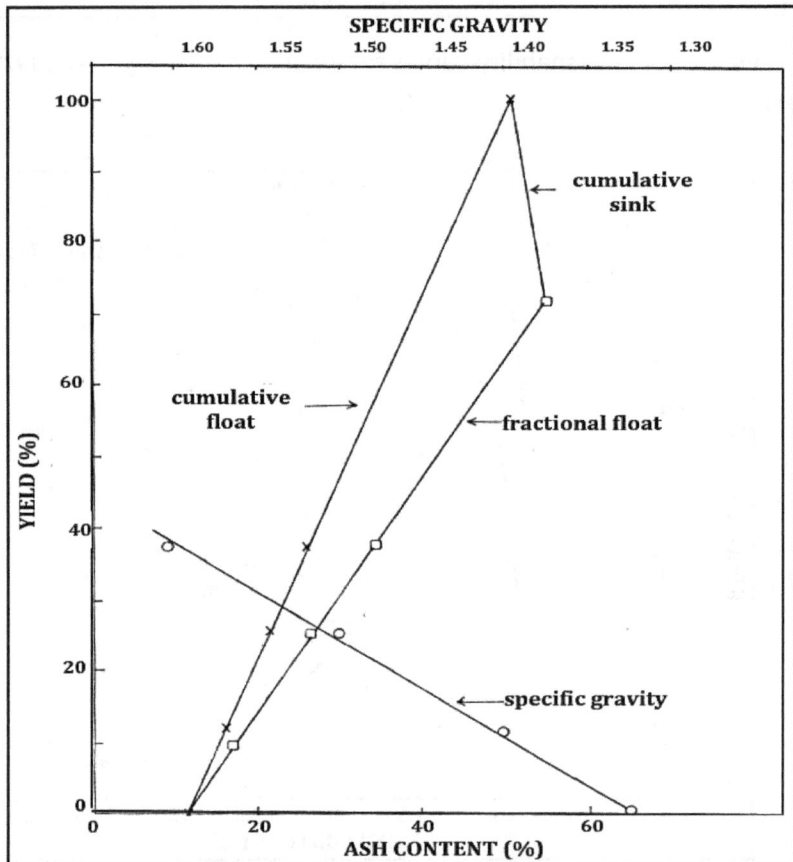

Figure 5.7 Washability curves for Obi/Lafia coal (seam 12). *(Afonja, 1979).*

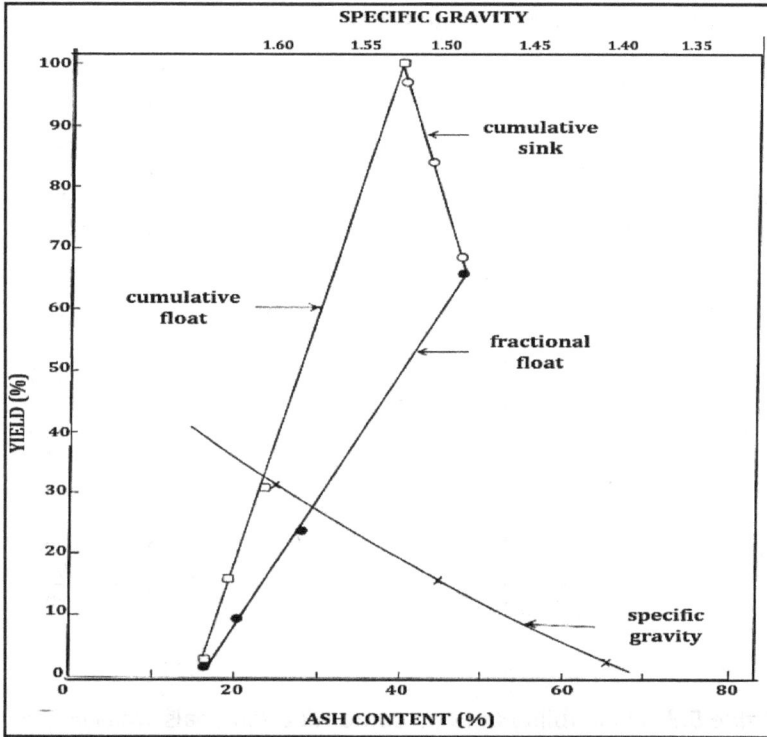

Figure 5.8 Washability curves for Obi/Lafia coal (seam 13). *(Afonja, 1979).*

Tables 5.2 and 5.3 compiled from two sources (Afonja, 1976, 1977, Estel, 1976) present further data on the washability of Nigerian coals. Data from both studies show that the ash content of Enugu and Orukpa coals can be reduced to an acceptable level below 10%, with good yield by washing in liquids of up to 1.5 specific gravity. At specific gravity of 1.25, the yield for Enugu coals is between 55% and 70%.The ash content of unwashed Orukpa and Okaba coals is within acceptable limits for most applications, in particular, steam raising and power generation. Provided the overburden is carefully removed prior to open cast mining, the coals are sufficiently clean for direct use. Obi/Lafia coal on the other hand has unacceptably high ash content and the results of washability tests presented here show its very poor response to washing. Even at specific gravity of 1.6, the yield is only between 24 and 31% at ash contents of between 24 and 43%. The reason for this is evident from the results of petrographic analysis of the coal discussed in Chapter 3. The ash is finely enmeshed in the coal matrix and fine grinding would be required to facilitate reduction.

Froth floatation tests in kerosene/turpentine solution were carried out on pulverized Obi/Lafia coal, in two size ranges of 1.7mm to + 850 microns and −600 microns to determine its response to this beneficiation process (Afonja, 1977). The response was too poor for the coarser size fraction. It is evident from the results presented in Table 5.4 that the coal's response to this process is also poor, with only 15 to 21% yield at 15 to 28% ash content respectively. In summary, Okaba and Orukpa deposits which can be mined by opencast technique may not require washing prior to use in power generation. Enugu coal has been used for power generation for many years without washing but if mining is mechanized it may be necessary to wash the coal. The coal responds well to washing but a washery will add about 10% to capital and operating cost requirements.

Table 5.2 Washability data on some Nigerian coals. *(Afonja, 1976, 1977a&b, 1979; Estel, 1976).*

COAL SOURCE	SPECIFIC GRAVITY	GRAIN SIZE (mm)	YIELD (%)	ASH (dry, %)
Enugu, Run-of-mine (ROM) (Onyeama/Okpara/Ribadu blend)		50-16	52.1	20.7
		16-1	39.7	20.5
		-1	8.2	30.3
Enugu, washed		50-16	81.8	9.6
		16-1	80.5	9.5
Yield referred to ROM Enugu coal			74.4	
Orukpa, (ROM)	1.5	50-16	80.0	6.5
		16-1	17.1	14.3
		-1	2.9	24.8
Orukpa, washed		50-16	99.7	
		16-1	92.0	
Yield referred to ROM Orukpa coal			95.7	
Okaba, ROM coal		50-16	77.9	10.1
		16-1	19.1	9.0
		-1	3.0	12.2
Okaba, washed		50-16	99.7	
		16-1	99.2	
Yield referred to ROM Okaba coal			96.6	

Table 5.3 Washability data on some Nigerian coals. *(Afonja,1976, 1977a&b, 1979).*

COAL SOURCE	SPECIFIC GRAVITY	YIELD (%)	ASH dry,%) RAW COAL	ASH (dry, %) CLEAN COAL
Enugu (Onyeama)	1.25	69.7		
	1.25-1.5	100	11.3	4.8
	1.5-1.6	-		5.7
Enugu (Okpara)	1.25	54.7		3.6
	1.25-1.5	98.0	12.1	7.1
	1.5-1.6	2.0		7.7
Enugu (Ribadu)	1.25	63.0		3.9
	1.25-1.5	98.4	11.6	6.3
	1.5-1.6	1.6		6.6
Orukpa	1.25	-		-
	1.25-1.5	100	8.3	8.8
	1.5-1.6	-		-
Obi/Lafia Seam 12	1.25	-		-
	1.25-1.5	6.6	57.8	20.6
	1.5-1.6	23.6		43.1
Obi/Lafia Seam 13	1.25	-		-
	1.25-1.5	16.0	44.2	19.4
	1.5-1.6	31.4		23.9

Table 5.4 Froth flotation of Obi/Lafia coal. *(Afonja, 1979).*

COAL SOURCE	YIELD (%)	INITIAL ASH (%)	INITIAL SULPHUR (dry, %)	FINAL ASH (dry, %)	FINAL SULPHUR (dry, %)
Obi/Lafia Seam 12	15.5	57.8	3.04	28.1	1.94
Obi/Lafia Seam 13	21.3	44.2	6.83	15.2	1.5

5.4.2   Desulphurization of Nigerian coals

There are several methods of removing sulphur from coal and selection depends on the ultimate use of the coal and the over-riding objective. For coal destined for combustion for steam raising and and power generation, the main goal is environmental protection and pyritic sulphur can be removed by pre-treatment/washing, air oxidation or chemical leaching. Sulphur in coal can also be removed at the combustion stage in fluidized beds by reacting it with calcium sulphate, or removed from the flue gases by scrubbing. Coal for use in metallurgical extraction must be desulphurized prior to carbonization, usually by washing. Physical cleaning can remove up to 90% pyritic sulphur, depending on the morphology. It is easier to remove when it is mainly in discrete lumps but when it is finely disseminated in the coal matrix it is more difficult to remove by washing unless it is finely ground.

As much as 40% of the coal could be lost in the 'sink' proportion of float and sink cleaning while cleaning finely ground coal requires froth flotation, a relatively expensive process. Also, the coal may be too fine for use without agglomeration. Sulphate sulphur is easily removed by washing since it is soluble in water but removal of organic sulphur requires thermal degradation of the coal. Apart from physical washing, chemical leaching is the most popular method of sulphur removal and various processes are available, using caustic solution, ferric chloride, ferric sulphate, or by oxidation to sulphate and subsequent removal by washing. Most chemical and thermal processes of sulphur removal reduce the heating value of coal due to part depletion of hydrogen and carbon in form of water and carbon dioxide respectively while the caking properties may be partially or completely destroyed.

All Nigerian coals with the exception of Obi/Lafia coal have low sulphur content (Afonja, 1975) (see Table 5.1). The sulphur in Enugu coal is predominantly organic but that of Obi/Lafia is mainly pyritic. In theory, organic sulphur in coal cannot be removed by washing, only drastic thermal degradation can reduce this type of sulphur. On the other hand, sulphate and pyritic sulphur of the right morphology should be easily removable by leaching. The sulphur in Enugu and Orukpa coals is mainly organic but the form has not been determined for Okaba coal. In any case the total sulphur content of the three deposits is less than 1% and well within acceptable limits for power generation. Furthermore, the proposed captive power plants will almost certainly feature fluidized-bed combustion and sulphur content can be reduced in-situ.

The sulphur in Obi/Lafia coal is mainly in pyritic form and, in theory should be removable by washing. However, washability tests on the coals did not result in a significant reduction of the sulphur contents unless the coal is pulverized (Afonja, 1979). This is due to the fact that the pyrites is not in the form of interbedded layers but is finely interspersed in the coal matrix and fine grinding would be required to free the grains. The coal would be unacceptable for use in metallurgical coking but could be used for power generation if the fluidized-bed combustion system is adopted. Moving grate combustion would require auxiliary flue cleaning installation to comply with environmental standards.

The sulphur content in obi/Lafia coal is not only high but it varies widely from 1 to 11%. This is due apparently to the structure of the coal bed which features randomly distributed sulphur compounds in layers adjacent to coal strata. The sulphur is 80-85% pyritic, in discrete lumps but also finely disseminated in the coal matrix (Figure 3.8). Microstructural examination of petrographic samples showed that the pyrites is present in the coal both as lumps (200-500 microns) and as very fine grains (2-8 microns) dispersed in the coal matrix. This suggests the presence of iron pyrites deposits in the vicinity of the coal bed and the diffusion into the coal matrix over millions of years.

The origin and forms of sulphur in coals have been studied in depth by Yurovsky (1960). On the basis of data on coals from many parts of the world, It was shown that sulphate sulphur, though very stable, never exceeds 0.1–0.2% in fresh coal. Higher values may be found in weathered coals due to the oxidation of pyrites to form iron sulphates. It was concluded also that pyritic sulphur usually constitutes by far the highest proportion of the total sulphur present in high sulphur coals. The data on Nigerian coals presented in Table 5.1 supports this conclusion. The data presented by Yurovsky also shows a very strong interrelationship between the pyritic and organic sulphur contents of coal (correlation coefficient = 0.99). Furthermore, there is a strong interdependence between the amount of sulphur in coal and the geochemical nature of the overburden. Low-sulphur coals tend to occur under clayey shales while an overburden of limestone indicates the presence of high sulphur coal. Available geological data on Nigerian coals support this postulation. The overburden of the coals in the Enugu-Okaba belt are mainly clayey while the overburden of Obi/Lafia coal seams is carbonaceous in nature.

The response of the Obi/Lafia coal to washing has been studied in depth by Afonja (1979). The focus was on seams 12 and 13 since they were the only ones considered exploitable economically. The sulphur content of 135 borehole samples ranged from 0.38 to 11.1%, with an average of 3.25%. This wide range of values suggests that the coal bed is intersparsed with discrete layers of sulphur compounds, predominantly pyrites and should in theory be removable by physical washing. However, extensive tests have shown that this is not economically feasible. This is due to the fact that a significant proportion is finely dispersed in the coal matrix. The coal responded poorly to float and sink beneficiation technique Figures 5.7 & 5.8). Although washing 0-12mm samples from seam 12 in liquid of 1.5 specific gravity reduced the ash content by about 70% and the sulphur by about 30%, the yield was less than 40%. The response of seam 13 samples was better, with cumulative float of 56%, 18% ash and 4.5% sulphur content at 1.5 specific gravity.

Float and sink washability tests on 6–12mm samples of Obi/Lafia coal resulted in a 30% reduction in the sulphur content at 1.55 specific gravity. However, the yield was less than 30% (Afonja, 1979). Froth flotation of minus 600 microns coal in 5:1 kerosene/turpentine mixture removed over 70% of the sulphur but the yield was only about 20%. Air oxidation of 0-3mm samples of the coal removed nearly 80% of the sulphur at temperatures 400°C and above but the agglomerating properties of the coal were completely destroyed. Froth flotation of minus 600 microns Obi/Lafia coal in kerosene/turpentine solution did not reduce the ash and sulphur to acceptable levels. Although the ash and sulphur contents of seam 13 coal were reduced by as much as 66% and 78% respectively, the yield was only about 21%.

The response of seam 12 was even poorer, with comparative figures of 51% and 36% at 16% yield. The addition of some sodium sulphate to the frothing solution resulted in slightly higher reduction in both ash and sulphur contents of both coals. Sulphur may be removed from coal by thermal treatment either through direct decomposition or by direct oxidation of pyrites. The kinetics of dissociation of pyrites is very temperature-dependent. Decomposition is slow and insignificant up to about 550°C and is virtually complete around 680°C. The oxidation rate of pyrites by oxygen is also a function of temperature. Reaction does not commence until about 450°C and becomes vigorous only at about 550°C. Most coals become plastic around 375°C and resolidify around 500°C. In effect, removal of sulphur from coal by thermal treatment is only feasible in a range of temperatures which is higher than the plastic range of coal. It is inevitable therefore that treated coal will lose most of its caking properties. This explains why the swelling Index of Obi/Lafia coal was reduced from 6.5 to zero as a result of thermal treatment.

The response of Obi/Lafia coal to air/steam treatment is very similar to that of air oxidation except that the temperature required to sustain a given reaction rate is slightly lower. Also, the effect of treatment on the agglomeration characteristics of the coal is less drastic. Sulphur removal is insignificant below about 200°C and is virtually complete at 350°C. However, even though these temperatures are below the softening point of most coals, they are sufficiently high to cause intense oxidation and consequent loss of caking properties. Particle size also has a significant influence on desulphurization. Fine particle size promotes intimate contact between the coal particles and the gaseous oxidant and it can be expected that the finer the particle size range the higher the desulphurization, other factors being kept constant. However, the finer the particle size the more difficult it is to utilize the desulphurized coal without agglomeration.

Steam/air oxidation of Obi/Lafia coal at 350°C and above removed 45–75% of the sulphur depending on the size consist, temperature and flow rate, the highest being for 0-3mm samples treated at 350°C and air flow rate of 5.0l/hr. The effect of this treatment on the caking properties was less severe compared with air oxidation, the swelling index having been reduced from 6.5 to 2. Desulphurization by chemical leaching with a ferric salt appears to be the most promising process for desulphurization of Obi/Lafia coal. (Afonja, 1979). Ferric sulphate is more effective than the chloride and important variables include particle size consist, ash content level, solution concentration, temperature and leaching time (Figures 5.9-5.16).

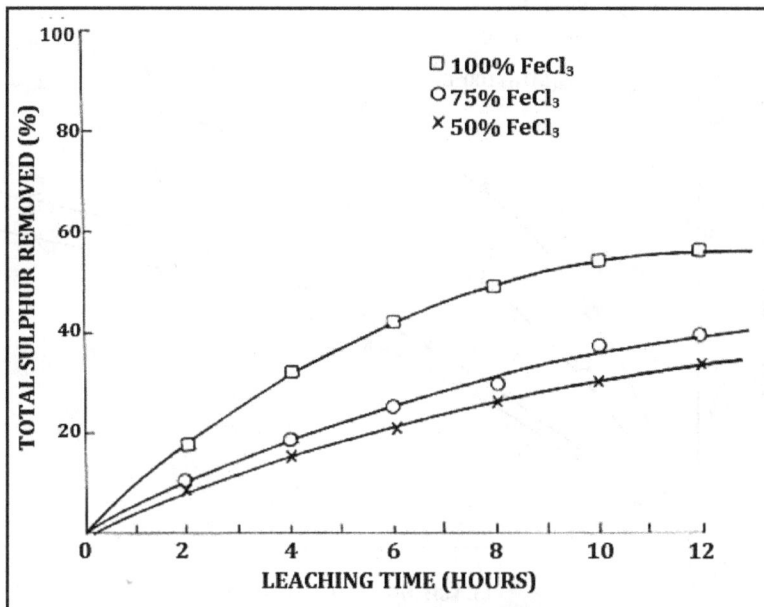

Figure 5.9 Effect of leachant concentration and time on desulphurization of Obi/Lafia coal, using ferric chloride solution at room temperature. Slurry concentration = 120g/litre, particle size distribution = 0-3mm. *(Afonja, 1979).*

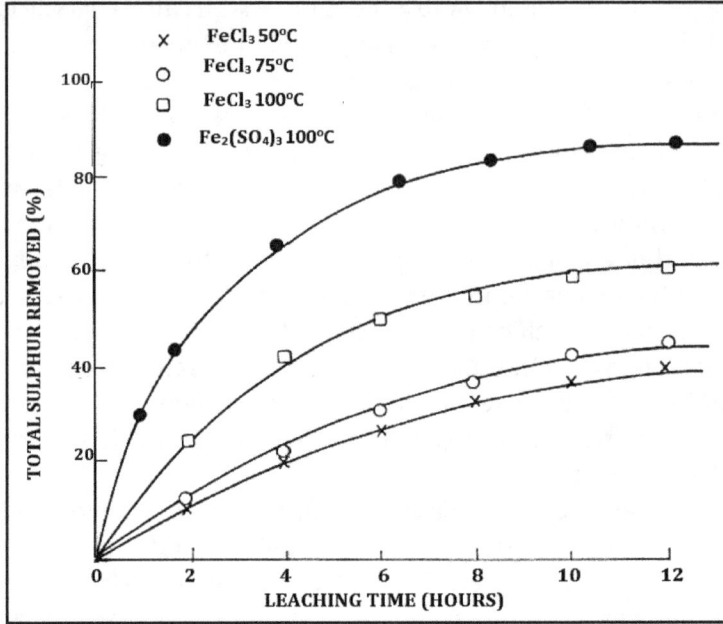

Figure 5.10 Effect of leaching time, temperature and type of solution on desulphurization of Obi/Lafia coal, Seam 13. Solution concentration = 0.5M, slurry concentration = 120g/l, particle size distribution = 0-3mm. *(Afonja, 1979).*

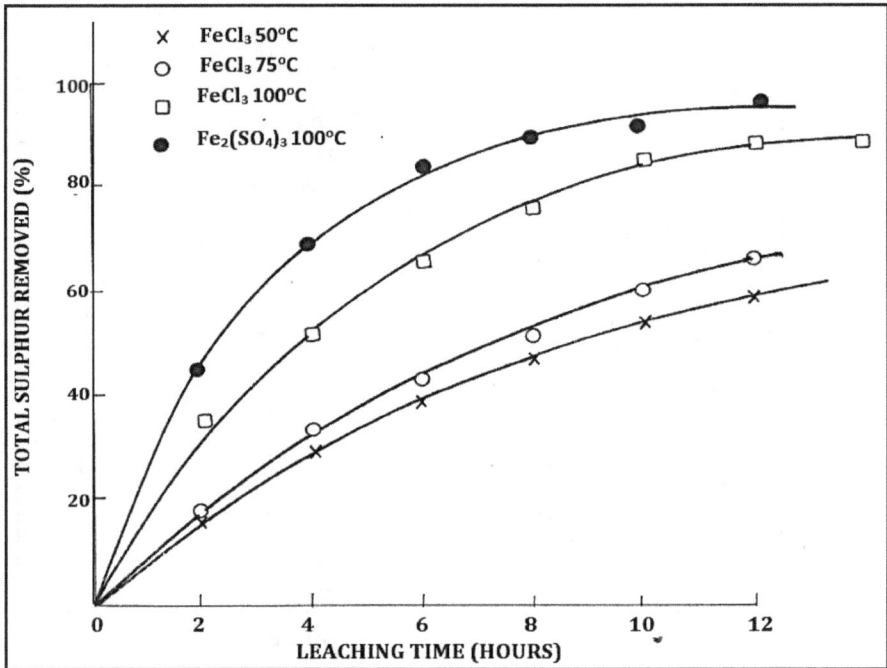

Figure 5.11 Effect of leaching time, temperature and type of solution on desulphurization of Obi/Lafia coal, Seam 13. Solution concentration = 1.0M, slurry concentration = 120g/l, particle size distribution = 0-3mm. *(Afonja, 1979).*

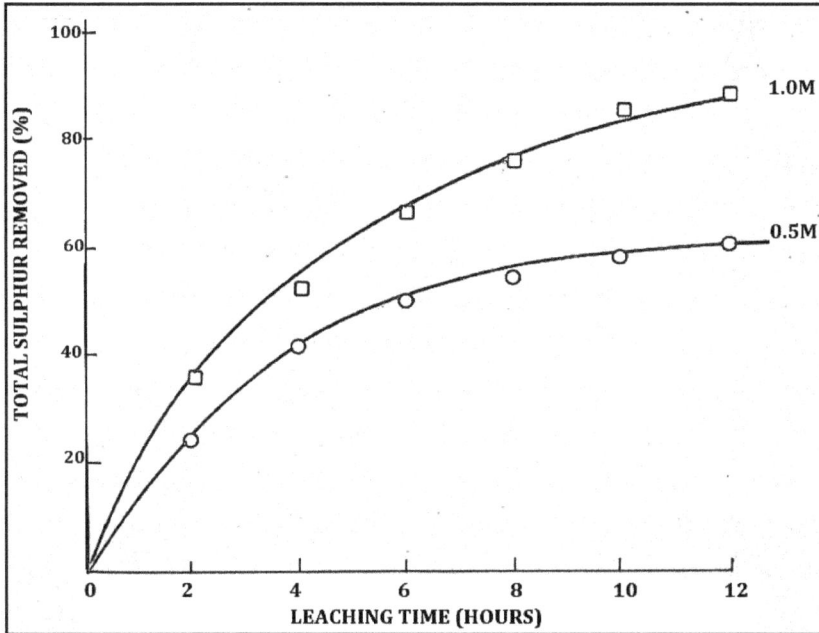

Figure 5.12 Effect of solution concentration and leaching time on desulphurization of Obi/Lafia coal, Seam 13 using ferric chloride solution, temperature = 100°C. Slurry concentration = 120g/litre, particle size distribution = 0-3mm. *(Afonja, 1979).*

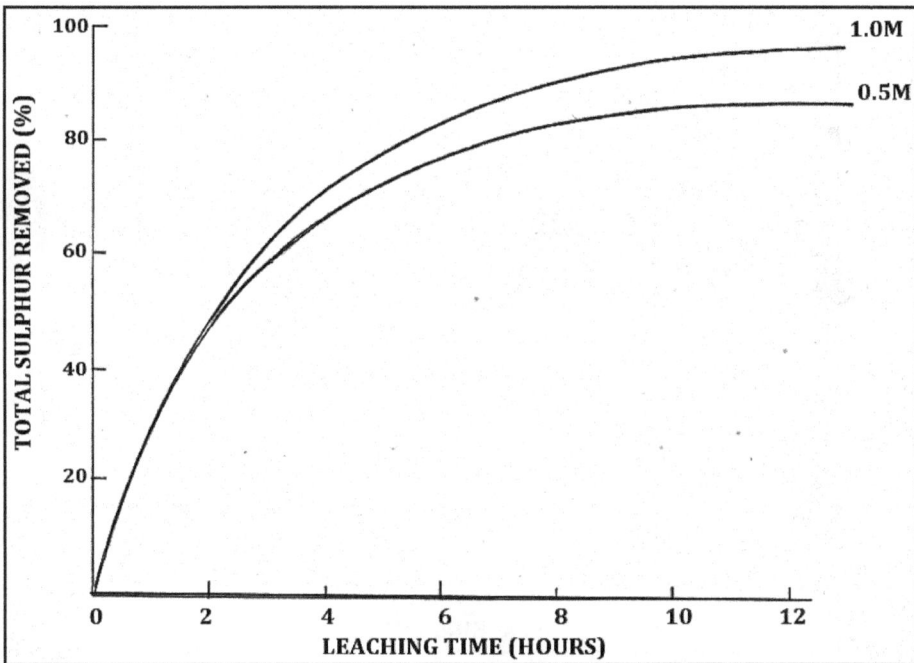

Figure 5.13 Effect of solution concentration and leaching time on desulphurization of Obi/Lafia coal, Seam 13 using ferric sulphate solution, temperature = 100°C, slurry concentration = 120g/litre, particle size distribution = 0-3mm. *(Afonja, 1979).*

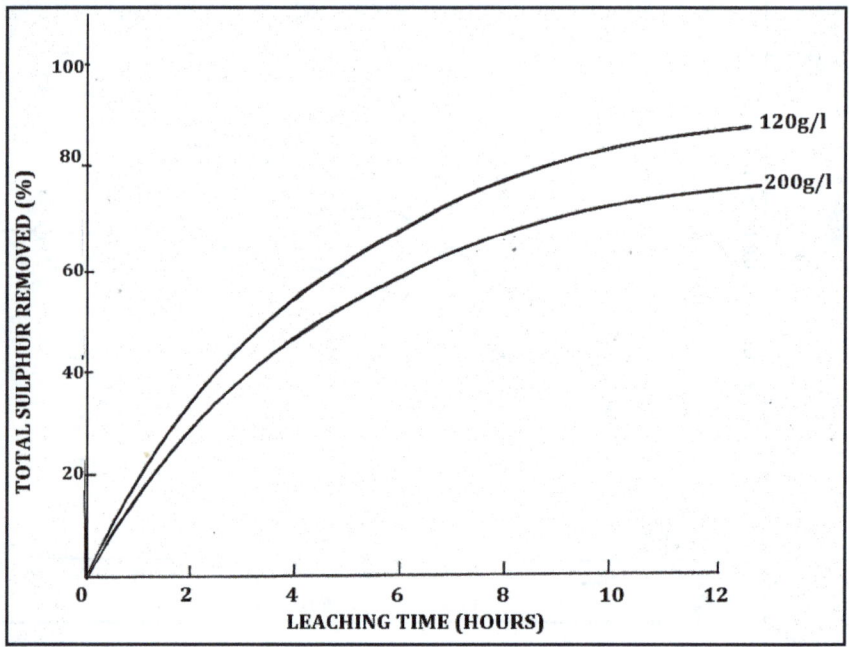

Figure 5.14 Effect of slurry concentration and leaching time on desulphurization of Obi/Lafia coal, Seam 13, using ferric chloride solution, concentration = 1.0M, temperature = 100°C, particle size distribution = 0-3mm. *(Afonja, 1979).*

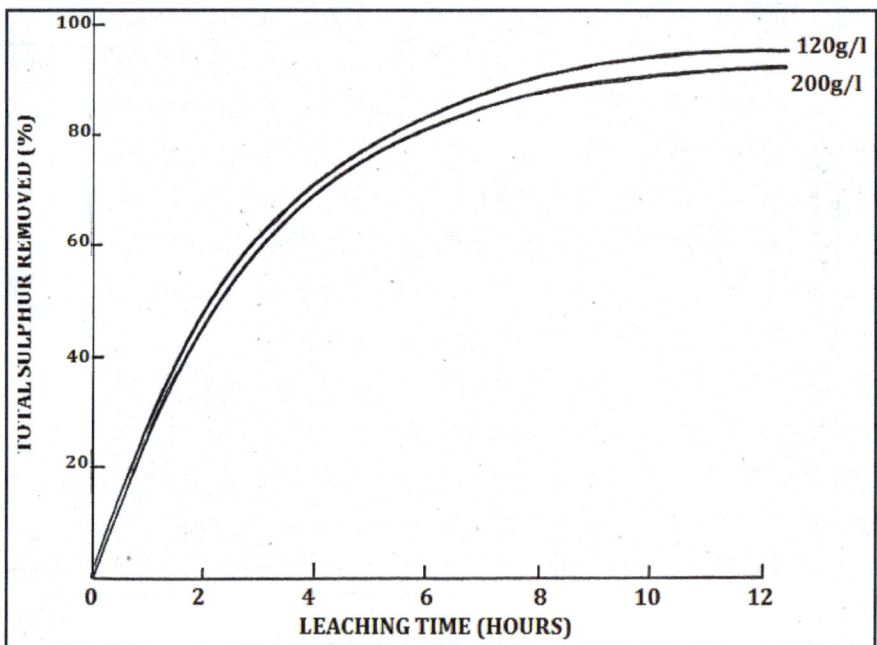

Figure 5.15 Effect of slurry concentration and leaching time on desulphurization of Obi/Lafia coal, Seal 13, using ferric sulphate solution, concentration = 1.0M, temperature = 100°C, particle size distribution = 0-3mm. *(Afonja, 1979).*

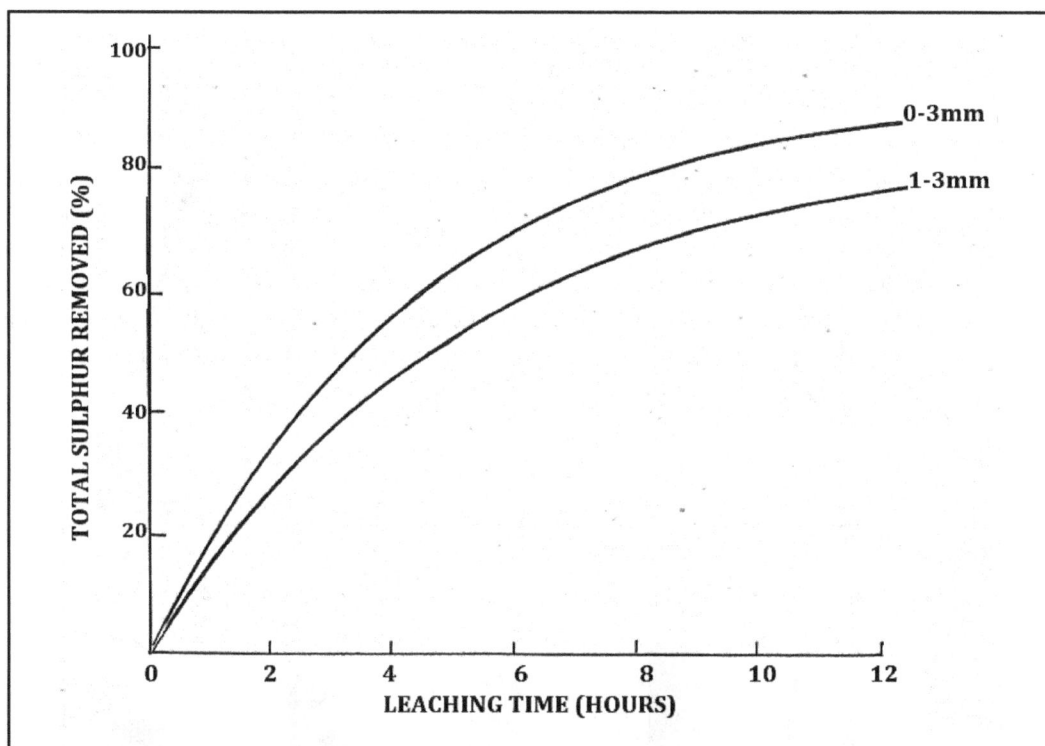

Figure 5.16   Effect of particle size distribution and leaching time on desulphurization of Obi/Lafia coal, Seam 13 in ferric chloride solution, solution concentration = 1.0M, temperature = 100°C, solution concentration = 120 g/l. *(Afonja, 1979).*

More recent work by Adeleke et al. on cleaning characteristics of Okaba and Obi/Lafia coals (Adeleke, 2010; Adeleke et al., 2011, 2013) involved single and multi-stage leaching of several size fractions with various basic and acidic reagents of varying concentrations at room and elevated temperatures. Tests were carried out at ambient and elevated temperatures. Samples of Okaba coal were obtained from seams 2 and 3 at depths of about 10m and 15m. The Obi/Lafia coal samples were obtained from outcrops at depths of about 1 meter. Hot leaching with aqueous sodium carbonate was found to be the most effective and economical method of reducing the ash and sulphur contents of the coals. The ash content of Okaba and Obi/Lafia coals were reduced by up to 13% and 36% respectively. Corresponding reductions in sulphur contents were 13% and 67% respectively (Figures 5.17 and 5.18). Particle size had a significant effect on both the reductions in ash and sulphur, with the highest yield, ash and sulphur removal when the coals were ground to - 250µm to + 125µm size range .

As expected the thermochemical quality of the coal improved. Volatile matter was reduced, the fixed carbon was higher and swelling index increased because of a decrease in ash-forming inorganic matter which contributes significantly to volatile matter content of coal. Reduced ash also means reduced infusible content of coal, hence the increase in thermoplasticity as indicated by the increase in swelling index. It should be noted however that the tests were carried out on outcrops and samples collected from seams 2 and 3 at depths of no more than about twenty metres implying that the samples may be highly weathered and unrepresentative of the bulk deposits. Furthermore, the seams had not been investigated previously because they were not considered viable. The sulphur content of the Obi/Lafia coal sample was only 0.91% compared with up to 12% on borehole samples (Afonja, 1976). It is probable that the hand-picked coal samples did not have much epigenetic intrusions. Furthermore, sulphur may be leached out of coal exposed to the atmosphere for a long time. It should be noted also that the investigation was carried out on a micro-scale and pilot plant tests on representative bulk samples would be required to confirm the above results. In any case, as mentioned earlier, Obi/Lafia coal is of low value in metallurgical applications, apart from the geological problems which may make mining uneconomical. All processes that can reduce the sulphur content to acceptable levels either have very low yield or destroy the caking properties. However, the deposit can be used for power generation provided that the plant is designed to process high-sulphur coals.

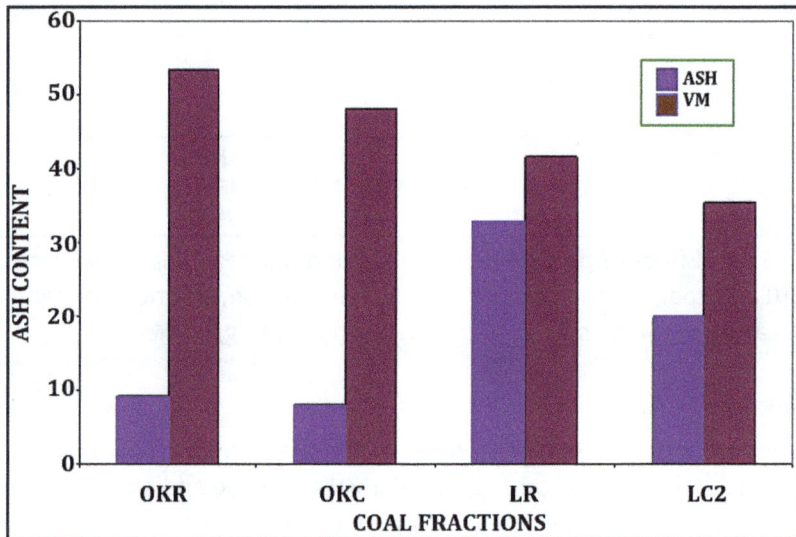

Figure 5.17 Reduction of ash and volatile matter content of Okaba and Obi/Lafia coals by agitation leaching with sodium carbonate. OKR = Okaba coal as received; OKC = Leached Okaba coal; LR = Lafia coal as received; LC2 = Leached Obi/Lafia coal *(Adeleke, 2010; Adeleke et al. et al, 2011).*

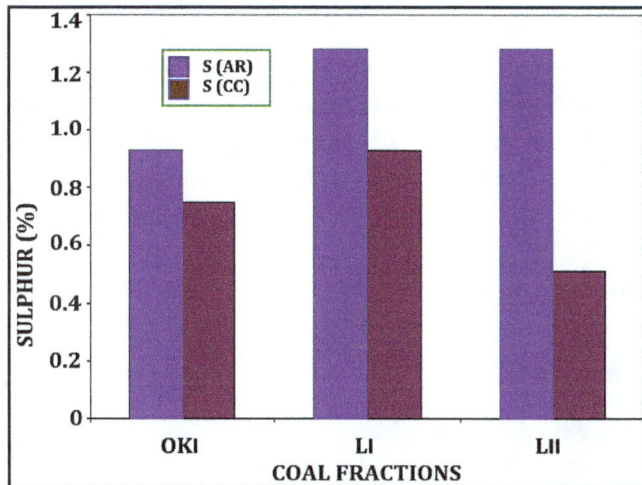

Figure 5.18   Reduction of sulphur content of Okaba and Obi/Lafia coals by agitation leaching with sodium carbonate. S(AR) = sulphur contents as received; SCC = sulphur contents of leached coals; OKI = Okaba coal leached with 1:40 sodium carbonate; LI = Obi/Lafia coal leached with 1:40 sodium carbonate; LII = ObiLafia coal leached with 1:20 sodium carbonate. *(Adeleke, Adeleke et al. et al, 2011)*

REFERENCES

Adeleke, A. O., Ibitoye, S. A., Afonja, A. A. and M. M. Chagga (2011) "Multistage caustic de-ashing of Nigerian Obi/Lafia coal." *Petroleum & Coal,* Vol. 54, No. 4, pp. 259-265

Adeleke. A. A., Ibitoye, S. O. and A. A. Afonja (2013). "Multistage caustic leaching desulphurization of a high sulphur coal." *Petroleum & Coal*, 55 (2), pp. 112-117.

Afonja, A. A. (1972). "An assessment of the coking quality of Nigerian coal seams." ENE/B/W/3. Report No. IFE/CHE/CP/1. For the Nigerian Steel Development Authority. (N.S.D.A.).

Afonja, A. A. (1975). "Chemical, Petrographic and Coking Studies of Enugu Coal." Nigerian Mining and Geology, 1975, 12, p. 40.

Afonja A. A. (1976). " Deashing and desulphurization of Nigerian coals. Research Report No. IFE/CHE/CP 06.

Afonja, A. A. (1977a). Washability charts for some Nigerian coals. Journal of Mining and Geology, 14(1), pp. 47-49.

Afonja, A. A. (1977b). Feasibility of cleaning Iva Mine coal : sink and float tests. Report No. IFE/CHE/CP/8 . For the Nigerian Coal Corporation.

Afonja, A. A. (1979). "Further Studies on the Obi/Lafia Coal Deposit (Phase II)." Research Report No. IFE/CHE/CP/11, Coal Research Laboratory, Ife.

Arthur D. Little Inc. (1965). Revitilizing the Enugu Coal Industry Through Large-Scale Carbonization. Duffryn Technical Services Ltd. Chemical Utilization Project Report to the Under Secretary of State for the colonies, London.

Estel (1976). "Investigations into possibilities of preparing and carbonizing Nigerian coals from Enugu District." Estel Exploration und Bergbau GmbH, Dusseldorf.

ROE GMBH (1982). "Feasibility Study of the Carbonization of Okaba Coal." Final Report prepared for the Nigerian Coal Corporation. Berlin, West Germany.

Ugwu, F. N. (1996). "Optimization of Nigerian Coal Production." *In Nigerian Coal: A Resource for Energy and Investments*. Okolo, C. and M. C. Mkpadi (Eds).Raw Materials Research and Development Council."Multistage caustic leaching desulphurization of a high sulphur coal."

Yurovsky, A. Z. (1960). "Sulphur of hard coals." *Ac. Sci. USSR, M.*, 1960, p. 295.

6 Conversion potentials of Nigerian coals: combustion and carbonization

6.1 INTRODUCTION

Coal is a very valuable natural resource because of the numerous potential areas of utilization, from heating and power generation to the production of high-value industrial chemicals and other products. The technical properties of Nigerian coals indicate a high utilization potential, in particular for power generation and as feedstock for the production of chemicals. Coal is burned mainly to raise steam for power generation or other industrial applications. Over 40% of the global annual utilization is for power generation. Another 40% or so is utilized for domestic and industrial heating, cement manufacture and other industrial processes and, the balance is carbonized for iron and steel production.

6.2 COMBUSTION PROPERTIES OF NIGERIAN COALS

Coal for combustion is either fed to the furnace in lumps or pulverized for more efficient handling and combustion. Any coal can be burned to produce useful heat, raise steam, or generate power but the process economics and environmental impact of coal burning have become very important variables in the last five decades or so and led to the development of a set of variables for assessing the relative potential of coal for combustion and indeed any application, notably

- Useful heat content per unit weight

- Useful heat content per unit cost of transportation

- Relative economics compared with other local, potential energy sources

- Moisture and ash content

- Ash chemistry

- Content of undesirable elements, in particular, sulphur

- Weathering characteristics

Some of the above variables are interrelated. For example, high moisture and ash imply low useful heat content and high moisture retention capability indicates that the coal may have a significant propensity to weather in storage, which in turn leads to deterioration in heat value and may cause coal to self-ignite in storage or during transportation. It should be noted however that the above variables are unimportant when coal is the only available or viable option.

The gross calorific values of Nigeria's mostly sub-bituminous and lignite coals range between 24 and 34 kJ/kg, very good values compared with coals being used for power generation in most countries, including developed countries. However, the high volatile matter means that a significant proportion of the heating value is in the gases and will be lost in the flue gases unless a secondary flue gas heat recovery system is incorporated in the design of the combustion system. The moisture and ash contents which are 3-16% and 4-26% respectively are within acceptable limits. It should be noted however that none of the coals has ever been subjected to cleaning processes on a commercial scale. Float and sink processing and drying will reduce these values to limits that are internationally acceptable Table 6.1). The ash chemistry of Enugu coal shows silica and alumina contents of 61-66% and 21-27% respectively, values that are typical of unwashed coals. The sulphur content of all Nigerian coals with the exception of Obi/Lafia coal are very low, less than 1%. Considering that other countries are utilizing coals with sulphur content as high as 10-12%, Nigerian coals have excellent potential for any utilization on the basis of sulphur content. Available data on the trace elements in Nigerian coals is presented in Table 6.2.

Table 6.1 Combustion properties of Nigerian coals. (Average values of data from three sources) *(Swardt, 1961; Afonja, 1974; ROE, 1982).*

Quality Parameter	Enugu	Inyi	Orukpa	Okaba	Obi/Lafia (washed)	Ogwashi
Ash (%, dry)	11.3	6.5	8.8	15.9	14.7	14.4
Vol.Mat.(%, daf)	46.7	-	51.9	43.9	31.7	64.6
Gross.Cal.Value (MJ/kg, daf)	33	24.3	31.2	30.3	34.2	30.5
Carbon (%, dry)	79.3	-	73.9	78	88.4	72.1
Hydrogen (%, dry)	5.1	-	4.6	6.1	5.6	6.6
Sulphur (%, dry)	0.75	1.47	0.46	0.7	1.53-6.52	0.78
Ash Analyss (%)						
SiO_2	61.66-65.9		63.7			
Fe_2O_3	7.09		2.4			
Al_2O_3	25.61-27.1		29.0			
MnO	-					
TiO_2	1.93		1.4			
CaO	1.53		1.3			
MgO	0.24					
K_2O	0.43		0.5			
Na_2O	0.70		0.6			
SO_3	0.26					
Grindability Index	423		405	235		

Trace elements are defined as elements present in coal in amounts of less than 1% by weight. Generally, trace elements are present in coal in amounts much less than 1%, and are often reported in parts-per-million (ppm) by weight in coal. Some of them are highly toxic, in particular, arsenic, mercury, lead and selenium. Despite the very small concentrations of trace elements in coals, the extensive worldwide utilization of coal results in emission of some of the elements in sufficient quantities to cause environmental concerns. For example, about 10 tonnes of lead is produced for every million tonnes for coal burned containing 10 ppm lead (wvnet.edu). Because of the potential effects of trace element pollutants from coal combustion and carbonization many environment regulation agencies worldwide have established acceptable limits for eleven elements classified as environmentally hazardous air pollutants (HAPs) in the Clean Air Act of 1990 (Davison, 1996) (Table 6.3).

Table 6.2 Trace elements in Nigerian coals. *(Udeagbara and Anusiobi, 2013).*

Element (wt%)	Ogwashi lignite	Orukpa sub-bituminous	Amansiodo sub-bituminous	Onyeama sub-bituminous
SO_3	0.056	0.058	0.085	0.058
Fe	0.236	0.365	0.635	0.536
Ca	0.005	0.003	0.006	0.008
Mg	0.024	0.009	0.012	0.013
Si	< 0.002	< 0.002	< 0.002	< 0.002
Na	0.005	0.007	0.008	0.015
K	0.019	0.018	0.019	0.010

Table 6.3 Acceptable limits for hazardous air pollutants (HAP). Clean Air Act of 1990. *(Davison, 1996).*

ELEMENT	CONCENTRATION (μg/g)	RANGE (μg/g)
Berryllium (Be)	2	0.1-15
Chromium (Cr)	20	0.5-60
Manganese (Mn)	70	5-300
Cobalt (Co)	5	0.5-30
Nickel (Ni)	20	0.5-50
Arsenic (Ar)	10	0.5-80
Selenium (Se)	1	0.2-10
Cadmium (Cd)	0.5	0.1-3
Antimony (Sb)	1	0.5-10
Mercury (Hg)	0.1	0.02-1
Lead (Pb)	40	2-80

The trace elements in a sub-bituminous coal and their behavior during combustion for power generation were studied by Querol et al., (1995). Thirty three elements including all the potentially environmentally hazardous elements were identified. Anhydrite calcium sulphate ($CaSO_4$) was found to absorb some of the elements effectively, in particular, arsenic, selenium and lead. The presence of trace elements in Nigerian coals has not been studied in any detail and available data does not include any of the HAPs (Table 6.3).

Enugu coal has good strength, a desirable property for stoker combustion. However, the grindability index is high and high specific energy will be required to pulverize the coal as a feed for a modern coal-fired power plant. Okaba coal is friable and decrepitates on prolonged exposure due to oxidation. Both coals ignite easily and have good combustion characteristics. These, coupled with low clinkering characteristics make the coals compare favourably with many foreign bituminous coals used for steam raising and power generation.

The relatively high ash, high volatile matter and low fixed carbon contents of Nigerian coals are disadvantages in coal combustion. However, many coals with much higher ash contents are in use all over the world. For example, South African coals have an average of 45% ash content, yet provide 75% of the country's primary energy needs, mostly electricity, considered to be the cheapest in the world. Over 75 million tonnes are also exported annually. All the other Nigerian coals have not been tested on any extensive basis. Although the contents of sodium and potassium compounds in the ash are low and the silica and alumina contents high, the high iron content of Enugu and Okaba coals causes depression of the ash fusion point and some clinkering.

6.3   CARBONIZATION PROPERTIES OF NIGERIAN COALS

Coal is heated in minimum air (carbonized) to expel the volatile matter content without burning the coal. The residue is nearly pure carbon (over 98%) with very high heating value. Carbonized coal (coke) has potential utility in many applications including iron and steel production, smokeless fuel briquettes, activated carbon, graphite, industrial diamond, etc. Depending on the thermochemical properties of a particular coal, the coke may be in a granular, or solid forms of varying strength and porosity. None of the Nigerian coals can transform into coke of metallurgical quality. All the coals with the exception of Obi/Lafia coal do not soften or go through any significant plastic phase (Enugu coal has slight thermoplastiticy). While this is an asset in coals used in power generation, such coals are unsuitable for metallurgical coke required for iron and steel production by the blast furnace technique. It is possible however to incorporate small quantities of the coals in a coking blend or for use as blast furnace fuel injection supplements. All the coals are also suitable for the production of smokeless fuel but the high sulphur content of Obi/Lafia coal would be a problem. However much of it can be removed if the coal is carbonized in a fluidized bed.

6.4   BLENDING CHARACTERISTICS OF NIGERIAN COALS

Although most Nigerian coals are unsuitable for metallurgical coking, the low ash and low sulphur contents make them suitable for incorporation in coking blends to moderate excessively caking coals. The coals may also be injected in blast furnace tuyeres in pulverized form as supplementary fuel to reduce coke rate. Extensive studies have been carried out to determine the extent to which the coals can be incorporated in a coking blend. The most important blending principle is that the behaviour of the blend must be as close as possible to that of a prime coking coal, in particular, fluidity, dilation, post solidification contraction, and swelling characteristics should be moderate and fall within a very narrow range typical of coking prime coking coals. Also, the softening range of the constituent coals of a good coking blend should overlap such that most components become fluid in the same temperature range, otherwise the coke formed will be weak and friable.

6.4.1   ROE coking tests on blends of foreign and Enugu coals

One of the earliest tests on the blending characteristics of Nigerian coals was carried out by (ROE, 1982). Coke char/breeze and carbo-pitch produced from Enugu and Okaba coals were blended with foreign coking coals in various proportions and coked in laboratory pilot coke ovens. The products were subjected to various coke quality tests. The ROE tests showed that a total of about 30% of Nigerian coals comprising 12% of high temperature coke breeze prepared from Enugu coal and about the same proportion of low-temperature char produced from Okaba coal, and approximately 8% of carbo-pitch produced from Okaba coal can be blended with high-volatile coking coal to produce coke of metallurgical quality.

6.4.2   Coking tests on foreign and Obi/Lafia coals blended with Enugu char

Extensive coking tests were carried out on blends of local and foreign coals by Afonja (1973, 1974, 1977, 1983, 1986, 1994, 1996a, 1996b, 1997), using the Miyazu and Simonis 'G' curves to determine the optimum blend composition. The locations of unblended Nigerian coals with respect to the coking windows are shown are shown in Figures 6.1 and 6.2 and data on several foreign coals were also plotted for comparison. An ideal coking coal/coal blend window is identified in both figures and coal blends whose overall properties fit into the window usually produce good coke. Low-temperature chars were prepared from Enugu coal in an externally heated fluidized bed at various temperatures (Afonja, 1981, 1983). The equipment is shown schematically in Figure 6.3 and described fully elsewhere (Afonja, 1983). Nitrogen was used as the fluidizing medium up to 600°C. High-temperature chars were prepared in a Carbolite pilot coking furnace at temperatures up to 1000°C. The chars were characterized by determining the residual volatile matter in accordance with BS 1016 Part 3. The dilation-contraction characteristics of compacted char pencils were determined using a high-temperature Linseis dilatometer between 20°C and 1000°C at a heating rate of 5°C. The residual plasticity of the chars was determined by the chloroform extract method. The effect of temperature and residence time on the rate of de-volatilization of Enugu coal is shown in Figure 6.4. De-gasification commenced at about 350°C and proceeded very rapidly up to about 600°C when about 90% of the volatile matter had been expelled. A change of residence time from 10 to 20 minutes did not have any significant effect on the rate of degasification. The contraction characteristics of chars prepared from Enugu and Obi/Lafia coals and a prime coking coal are shown in Figure 6.5 and Figure 6.6 shows the contraction coefficients as a function of temperature. The coefficients were calculated from Equation 6.1.

$$\delta = 1/l_o . dl/d\theta \hspace{4cm} [6.1]$$

where δ is the contraction coefficient, l_o is the length of the sample at resolidification, and $dl/d\vartheta$ is the slope of the contraction at temperature ϑ derived from Figure 6.5.

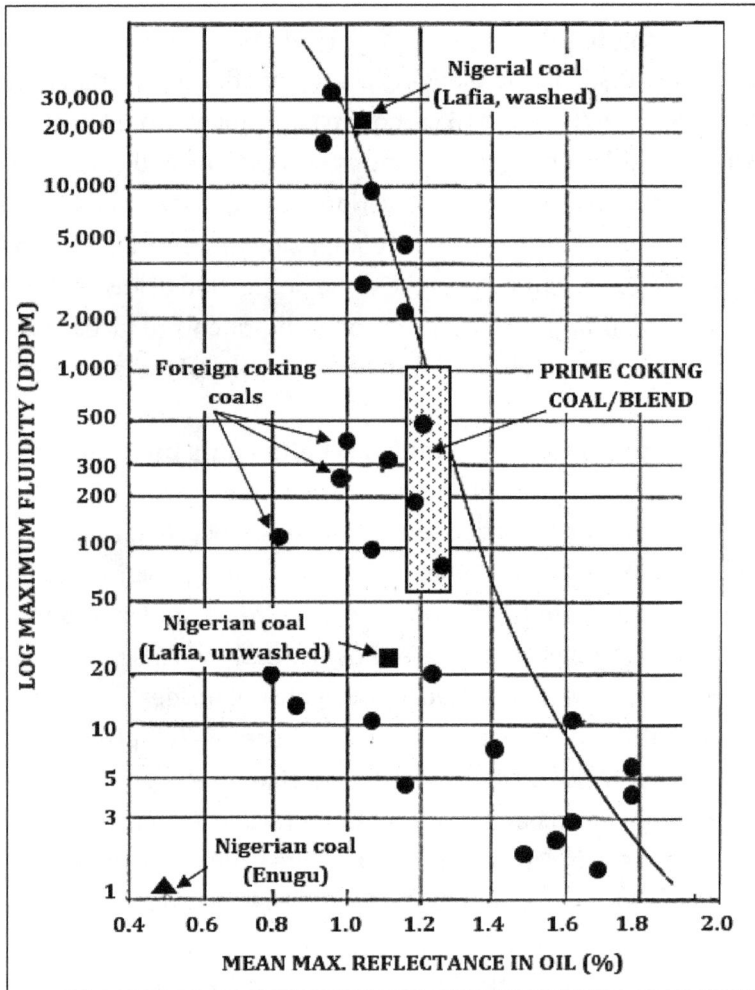

Figure 6.1 The Miyazu curve for coal blend design showing Nigerian coals in relation to the ideal coking blend window. (*Afonja, 1974*).

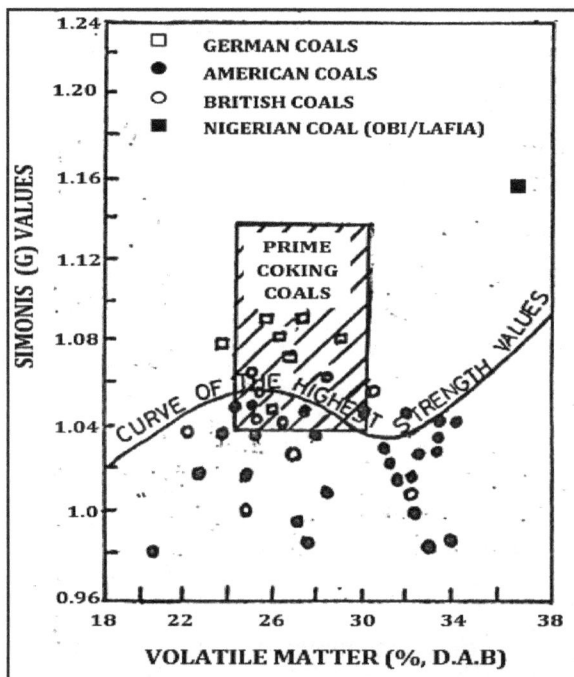

Figure 6.2   Simonis 'G' values for some Nigerian and foreign coals. *(Afonja, 1985).*

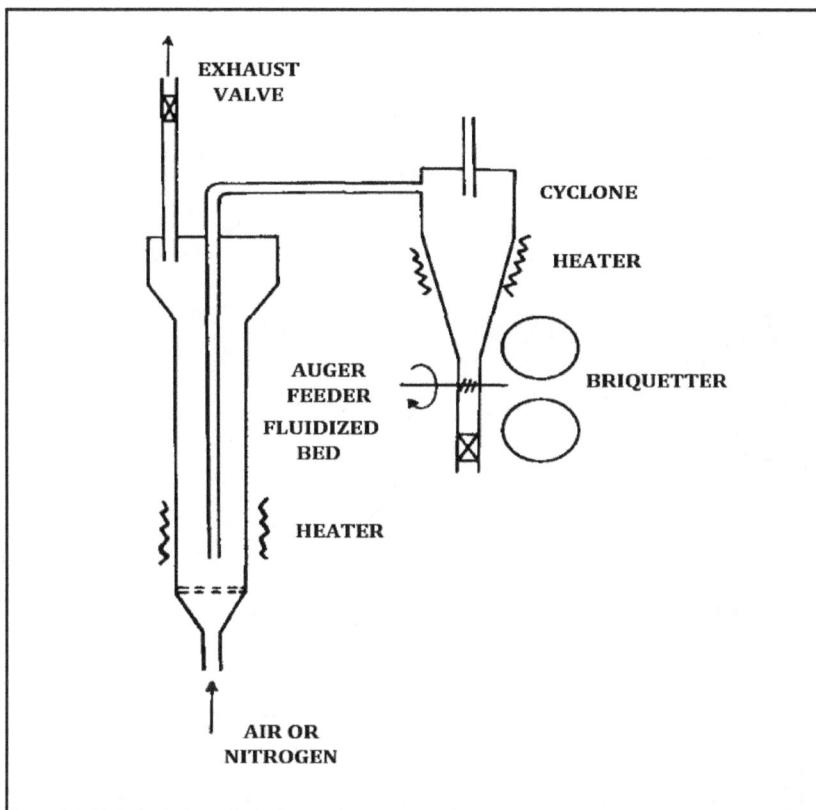

Figure 6.3 Schematic diagram of the apparatus for the production of char from Nigerian coals. *(Afonja, 1981).*

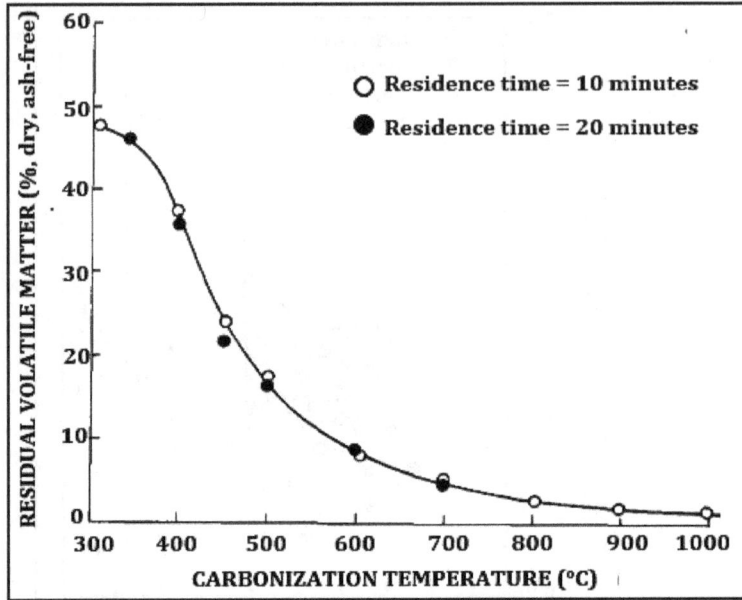

Figure 6.4 Effect of temperature and residence time on the rate of devolatilization of Enugu sub-bituminous coal. *(Afonja 1981).*

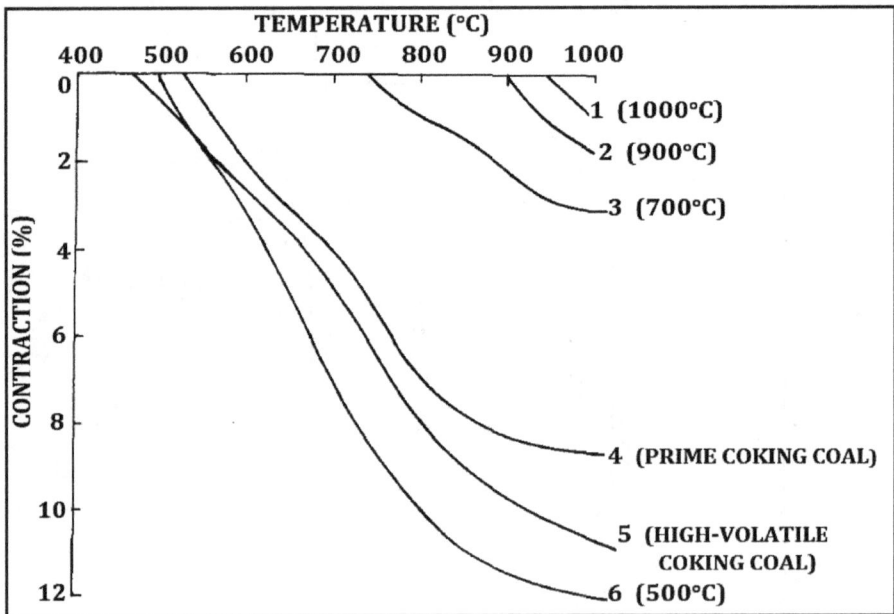

Figure 6.5 Contraction behaviour of coals and chars in the temperature range 400 to 1000⁰C. (1), (2), (3), (6) were sub-bituminous chars prepared from Enugu coal at the temperatures shown; (4) was a prime coking coal and (5) was Obi/Lafia coal *(Afonja 1983a&b).*

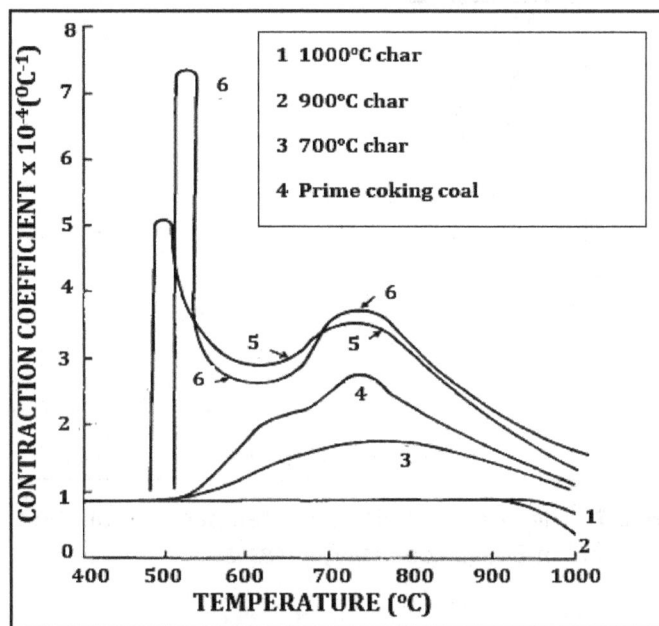

Figure 6.6 Contraction coefficients derived from Figure 6.5. *(Afonja 1983a&b).*

Two variables are important in the design of caking coal-char blends: the contraction characteristics of the blend components: the contraction characteristics of the blend components should be similar and the char should have optimal residual plasticity. The primary interest was to determine the feasibility of moderating the excessive caking properties of Obi/Lafia coal with char prepared from Enugu coal. Although the contraction behaviour of the two (samples 5 and 6 in Figure 6.6) were similar, there was no correspondence between the first peak coefficients but they were sufficiently close to be compatible. The residual plasticity of the sub-bituminous char as characterized by the chloroform extract is shown in Figure 6.7. It is clear from Figures 6.4 to 6.7 that char prepared from Enugu coal around 500°C has the best characteristics as a coking blend component in terms of residual volatile matter (34% - 26%), contraction characteristics, residual plasticity (1.2 - 0.8%), and potential compatibility with other components of a coking blend.

Coking tests were carried out in a 7kg Carbolite pilot coke oven on samples of a prime coking coal or Obi/Lafia coal blended with Enugu char prepared at various temperatures between 450°C and 900°C (Afonja, 1986). Previous studies had shown that partial agglomeration of a coking blend can improve the coking characteristics of the blend significantly (Nakamura et al, 1977). It is believed that the main effect of agglomeration is the increase in bulk density of the blend which generally results in improved coke properties, hence the various commercial practices which include pre-drying or stamping of the charge or mixing with oil. Blends of prime coking coal or washed Obi/Lafia coal with low-temperature and high-temperature Enugu chars were prepared and a proportion of each blend, between 5% and 30% was compacted on a briquetting machine, using soft pitch extracted from the

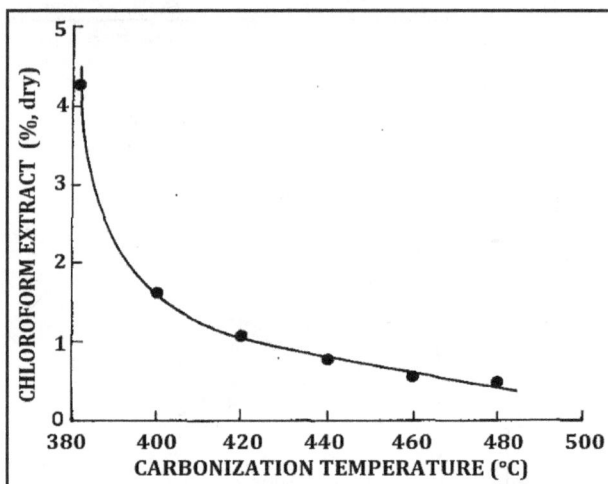

Figure 6.7 The effect of fluidization temperature on the chloroform extract yield of Enugu coal. *(Afonja, 1983a&b)*

flue gases of the fluidized-bed carbonizer. The blends were carbonized in the coke oven at 900°C for approximately 3 hours, quenched with cold water on a grate, dried and tested for strength and abrasion resistance in a Micum drum (BS 1016 Part 13). The effect of addition of low temperature char on the dilation of Obi/Lafia coal is shown in Figure 6.8. Addition of char moderated the dilation properties significantly, the proportion for optimum dilation being 10 - 20%. The results of coking tests on blends of a prime coking and Obi/Lafia coals with Enugu char are shown in Figures 6.9 and 6.10 respectively.

All additives had a detrimental effect on both the strength (M_{40}) and abrasion resistance (M_{10}) of the coke produced, the magnitude being highest when the blend contained no agglomerated briquettes. Addition of about 25% of minus 2mm low-temperature char to Obi/Lafia coal with 30% partial agglomeration increased the coke strength index (M_{40}) from 51 to 74%, or to 85% if the char had a narrow size range of <2mm to >600 microns (Figure 6.10). The abrasion index (M_{10}) was also within the acceptable range of 6 - 10%, with closely sized char producing better values. High-temperature char additive also had a positive effect on both the M_{40} and M_{10} indices but to a lesser extent compared with low temperature char. The optimum content was only 10 to 15% compared with 25 to 30% for low temperature char. Previous work by ROE (1982) had shown that char prepared from Enugu or Okaba coals as additives to a coking blend should be pulverized as fine as possible. Further tests were run to establish the optimum size range for char prepared from Enugu coal at 500°C in three size ranges up to 600 microns. Coking was carried out without partial agglomeration.

The results presented in Figure 6.11 show that pulverization of the char to below 100 microns produced coke with good strength and abrasion values but higher particle size ranges caused rapid deterioration in both indices. Samples of coal briquettes and coke produced are shown in Figures 6.12 and 6.13.

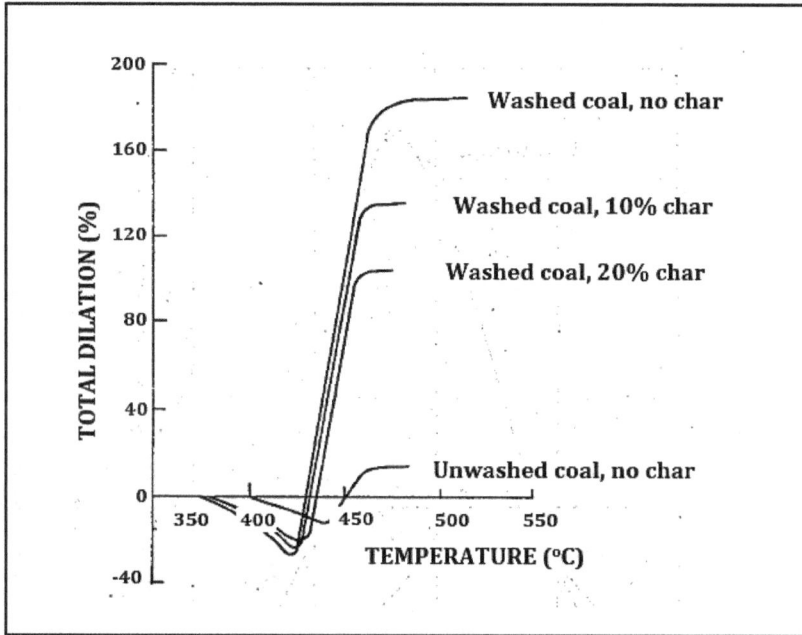

Figure 6.8 Dilation curves for Obi/Lafia coal with and without low-temperature char addition. *(Afonja 1983a&b).*

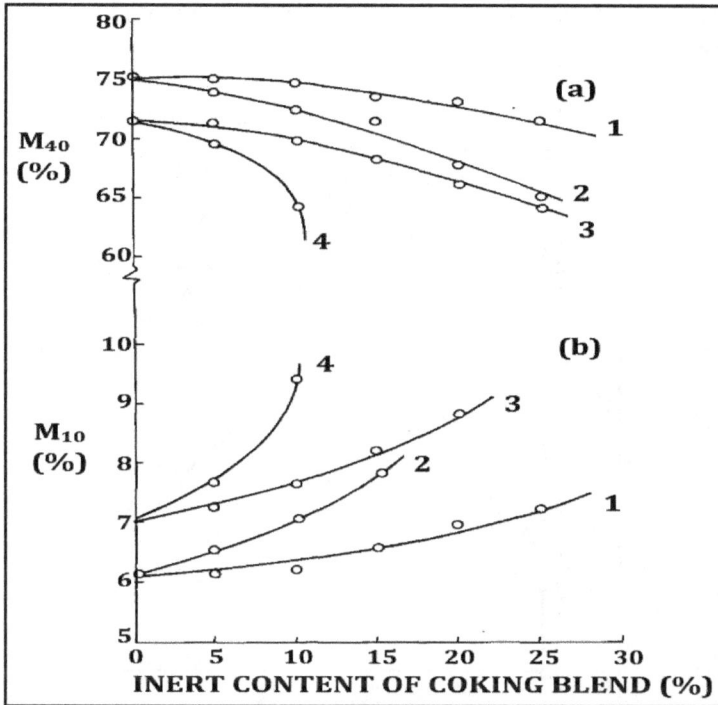

Figure 6.9 The effect of Enugu sub-bituminous coal or char additives on the strength and abrasion resistance of coke made from a prime coking coal. (1) low-temperature Enugu char additive with agglomeration of 30% of blend; (2) Enugu coal additive with agglomeration of 30% of blend; (3) low-temperature Enugu char additive without agglomeration; (4) Enugu coal additive without agglomeration. *(Afonja, 1983a&b).*

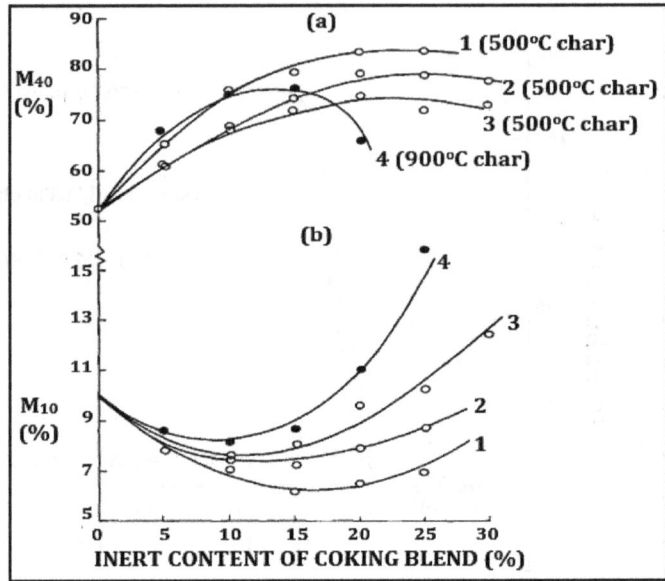

Figure 6.10 The effect of Enugu sub-bituminous coal or char additives on the strength and abrasion resistance of coke made from washed Obi/Lafia high-volatile coal.
(1) < 2mm to > 600 microns low-temperature Enugu char additive with agglomeration of 30% of blend; (2) < 2mm low-temperature Enugu char additive with agglomeration; (3) < 2mm high-temperature Enugu char additive without agglomeration; (4) < 2mm high-temperature Enugu char additive without agglomeration. *(Afonja, 1983 a&b).*

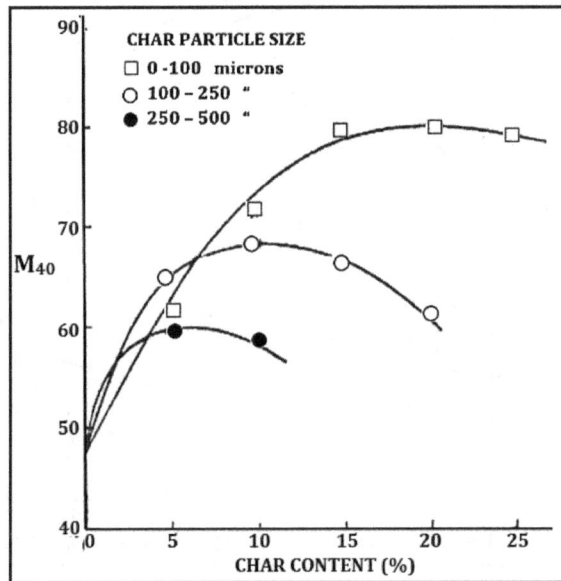

Figure 6.11 Effect of char particle size range on coke produced from washed Obi/Lafia coal-Enugu low temperature char blends without agglomeration. *(Afonja, 1979, 1983a&b).*

Figure 6.12 Sample of agglomerated washed Obi/Lafia coal-Enugu low temperature char coking blend. *(Afonja, 1983 a&b).*

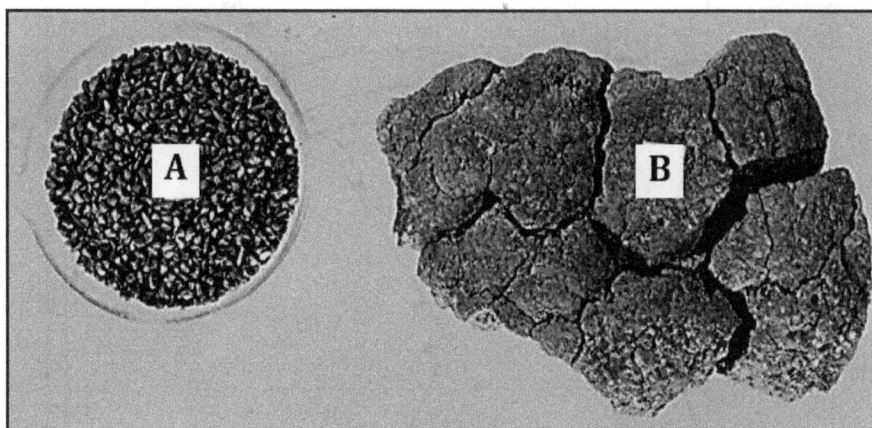

Figure 6.13 Samples of A) coking blend prior to agglomeration and (B) coke produced from washed Obi/Lafia coal blended with 25% low temperature char prepared from Enugu coal, with agglomeration of 30% of the blend. *(Afonja, 1983 a&b).*

The influence of partial agglomeration on the bulk density of Obi/Lafia coal-Enugu char blend is shown in Figure 6.14. The bulk density increased approximately linearly with increasing proportion of briquettes up to 30% partial agglomeration. Addition of 30% briquettes increases the bulk density of the charge by about 20%. The influence of bulk density on the M_{40} index of the coal-char blend is shown in Figure 6.15. For the same bulk density and char content, partially agglomerated blends produced stronger coke than homogeneous blends. Around 20-30% partial agglomeration produces coke of good strength, above 80%.

It is clear from the extensive data presented above that addition of char has a negative effect on coke produced from prime coking coals irrespective of proportion. On the other hand, washed Obi/Lafia coal blended with Enugu low-temperature char produced good quality coke if at least 20-30% of the blend is agglomerated. It should be noted however that this conclusion is based on coke strength and abrasion index only. The high sulphur content of the Obi/Lafia coal in spite of washing would be a constraint. The results of the coking tests show that the residual volatile matter is a primary variable for characterizing char used as coking blend constituents (see also Gregory and Horton 1970). High values of residual volatile matter in char may cause violent degasification during carbonization, resulting in the reduction of coke strength. On the other hand, char containing low residual volatile matter has a similar effect on coke strength since particles are hard and do not fuse during coking.

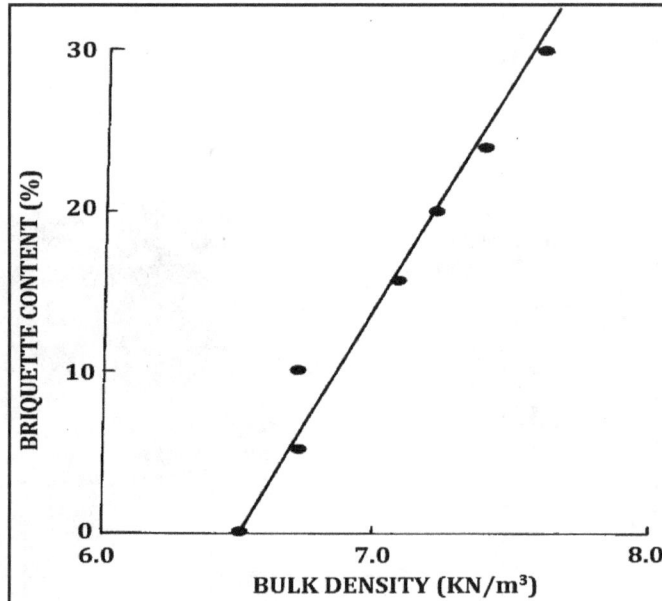

Figure 6.14 The relationship between the bulk density of washed Obi/Lafia coal and low-temperature Enugu char, and the proportion of agglomerates. All blends contained 25% char prepared at 450°C. *(Afonja, 1983a&b).*

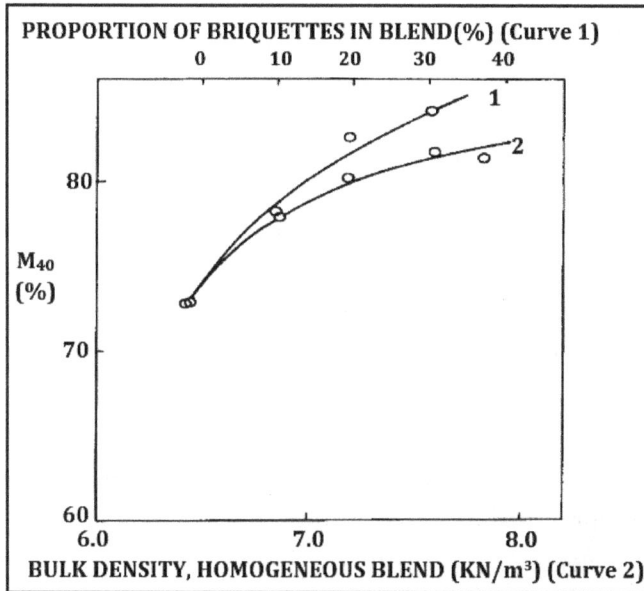

Figure 6.15 The effect of partial agglomeration on the strength and abrasion resistance of coke made from blends of Obi/Lafia coal and low-temperature Enugu char. (1) blends containing 0 to 30% agglomerates and 25% char; (2) homogeneous blends. (Bulk densities were varied by the addition of closely sized char of different size ranges. *(Afonja 1983a&b).*

The optimum value of the residual volatile matter appears to depend on the nature of the coal, but the results of previous studies suggest a range between 18 and 22% on dry, ash-free basis (Gregory 1966, 1971). This is near the range of low-volatile coals which it replaces in a coking blend. Furthermore, most coals soften in the temperature range in which tar is released and retention of some of the tar (residual plasticity) is believed to enhance the quality of the char as a coking blend. The values established for Enugu char were significantly higher - a residual volatile matter range of 25 - 34% was required to obtain an optimum residual plasticity of 0.8 - 1.2% as determined by the chloroform extract (see Figures 6.4 and 6.7). This corresponds to a fluidization temperature range of 420 - 440°C. In theory, high residual plasticity of blend component char should enhance coke strength since it should promote the fusion between the char and other components of the blend on carbonization. However, initial tests on blends of washed Obi/Lafia coal and high volatile Enugu char with residual plasticity of 1.2% (34% volatile matter) produced very weak coke with M_{40} below 20%. This was due apparently to the negative effect of rapid degasification on coke strength. Subsequent preliminary coking tests on the coal blended with char prepared at 500°C and containing about 25% residual volatile matter yielded good quality coke, hence the char was used for most of the blending tests.

A change of residence time from 10 to 20 minutes did not have any significant effect on the rate of degasification or on the chloroform extract. This contradicts the results of earlier work on medium volatile coal which showed a significant effect (Gregory and Pierce, 1977). It appears that the difference was due to the relatively low devolatilization rate of medium volatile coal compared with Enugu high-volatile sub-bituminous coal. The work of Kirov and Peck (1970) also showed that, for high volatile coals, the effect of residence time was secondary to that of carbonization temperature. Their results also showed that the relationship between the residual volatile matter and carbonization temperature was dependent upon the initial volatile matter of the coal especially at temperatures below about 700°C.

The lack of correspondence between the peak coefficients of the low-temperature char and the base coals (Figure 6.6) may cause cracks at the char-coal particle boundaries, thus weakening the resultant coke. The high temperature chars have negligible contraction coefficients, and hence should significantly reduce the peak coefficients of the base coals and fissuring of the coke. In effect, the addition of high-temperature char to caking coal, particularly high-volatile coal should improve the strength of the coke. However, this positive effect may be counteracted by the negative effect of the infusible char particles on the overall plasticity of the blend. Even in the plastic range the hard particles may act as crack nucleators and promote fissuring which in turn results in weak coke. This may explain the rapid deterioration of coke strength when more than 10-15% high temperature char was added to Obi/Lafia coal (Figure 6.10).

The influence of bulk density on the M_{40} index of the Obi/Lafia coal-char blend raises some doubt about the common assumption that the effect of partial agglomeration on coke strength is solely attributable to increases in bulk density (Kohno, 1988). For the same bulk density and char content, partially agglomerated blend produced stronger coke than homogeneous blends (Figure 6.14). One feasible explanation is the fact that, apart from bulk density effects, partial agglomeration promotes a more intimate interaction between the caking components and the inert constituents of the blend, thereby enhancing fusion in the plastic phase. This is supported by the work of Leeder and Price (1977) which showed that the addition of small, spherical agglomerates to a coking blend improved coke quality without increasing bulk density. The positive effect of partial agglomeration of coking blend on coke quality diminishes with increase in the proportion of agglomerates.

In contrast with the 30% optimum agglomerate proportion established by Afonja (1983), Nakamura et al. (1977) reported an improvement in coke strength with increase in briquette proportion up to about 50%, after which the effect diminished. They also observed a corresponding decrease in bulk density. The conclusion that can be drawn at this point is that the optimum briquette proportion may depend on the ranks and types of coals in the blend, particle size distribution, method of agglomeration, and so on. It should be noted however that, although partial agglomeration of coking blend enhances coke quality, the increased bulk density could increase the oven wall pressure excessively, thereby damaging the refractory lining and also causing an increase in the pushing load considerably. Nakamura et al. have suggested a maximum of 30 to 40% to avoid this problem.

An investigation on the suitability of Enugu and Obi/Lafia coals as blend components of coking blends was reported by Ndaji and Marsh (1985). Their results showed that it is possible to use Obi/Lafia coal as a bridging coal between Enugu coal and a prime coking coal in a ternary blend for making blast furnace coke. Further work by the same authors (1986) also showed that the excessive ash and sulphur contents of Obi/Lafia coal could be moderated to acceptable levels by incorporating the coal in a ternary blend comprising Obi/Lafia coal, Enugu coal and a prime coking coal.

REFERENCES

Afonja, A. A. (1973). "Preliminary studies on the briquetting of Nigerian coals. " *Institute for Briquetting and Agglomeration*, pp. 265-271.

Afonja, A. A. (1974). "Evaluation of the coking characteristics of some Nigerian coals by thermogravimetry." *West African Journal of Science*, 19(2), pp. 171-74.

Afonja, A. A., Lau, I. T. and Leeder, W. (1977). "A formed-coke pilot plant facility." *Institute for Briquetting and Agglomeration,* 14, 239 - 250.

Afonja. A. A. (1981). Fluidized-bed Carbonization of Nigerian Coals. *Proc. Nig. Soc. of Chemical Engineers*, p. 50.

Afonja, A. A. (1983a). "Optimization of the utilization of a sub-bituminous coal in conventional and form-coking." *Fuel Processing Tech.*, 7, 293-310.

Afonja, A. A. (1983b). "A study of the potential of Nigerian coals for metallurgical coking." *Proc. 1st National conference on steel*, Ovwian-Aladja, April 18-20, pp. 51-95.

Afonja, A. A. (1986). "A Study of the Potential of Low Grade Coals for the Production of Blast Furnace Coke. " *Proc. Int. Iron & Steel Congress,* Washington D. C. pp. 239-246.

Afonja. A. A. (1994). "Coal Blending for Blast Furnace Coke Production: The Effect of Blend Composition and Pre-Treatment on Coke Quality. " *Proceedings of the International Congress on Metallurgy and Materials Technology*, Sao Paulo, Brazil, Oct. 9-14.

Afonja. A. A. (1996a). "Coal blend design for blast furnace production." *Carbon '96*. The British Carbon Group, 173-175.

Afonja, A. A. (1996b). "Production of metallurgical coke from non-coking coals." In *Nigerian Coal: A Resource for Energy and Investments*. Ed. H. Okolo, RMRDC.

Davison, R. M. (1996). "Trace elements in coal." *Energeia,* Vol. 7, No. 3.

Gregory, D. H. (1966). "formed-coke from high volatile coals." Congress Int. de Chaleroi; A6.

Gregory, D. H. and A. E. Horton (1970). "Contraction modifies in the manufacture of coke." *J. Inst. of Fuel*, V. 43, pp. 389-96.

Gregory, D. H. (1971). " Contraction Modifiers." *Yearbook, Coke Oven Managers Association*, pp. 174-190.

Gregory, D. H. and T. J. Pierce (1977). "UK views on blend composition and preparation." *Proc. Conference on Coal, Coke and the Blast Furnace*, Middlesborough, 1977, p. 107.

Kirov, N. Y. and M. A. Peck (1970). " Characteristics of chars from fluidized-bed coal carbon-ization." *Fuel*, Volume 49, Issue 4, Pp. 375-394.

Kohno, T. (1988). "Substitution and reduction of coking coal in cokemaking." *Coal and Coke.* Int. Iron & Steel Inst., Technical Session, Committee of Technology. Brussels, April, 1988.

Leeder, W. and Price, J. (1977). The effect of partial agglomeration of coke-oven charges upon coke quality. *Proc. Of the 15ᵗʰ Biennial Conference of the International Briquetting Association*, Montreal, Canada, p. 213.

Nakamura, N., Togino, Y. and T. Tateoka (1977). "Behaviour of coke in large blast furnace." *Coal, Coke and Blast Furnace.* The Metals Society, London, pp. 1-18.

Ndaji, F. E. and Marsh, H. (1985). An Assessment of the suitability of Lafia and Enugu Coals as Component of Coking Blends in the Manufacture of Blast Furnace Coke by the Conventional Coke Oven Method. *Proceedings, Nigerian Metallurgical Society*, Jos, Nigeria.

Ndaji, F. E. and Marsh, H. (1986). Variation Pattern of Sulphur and Ash contents of Metallurgical Cokes with Changes in the Composition of Coal Blends: A Study using Lafia Coal, Enugu Coal and Imported Coking coal., Nigerian Met. Society, Ajaokuta, Nigeria.

Querol, X., Fernandez-Turiel, J. and A. Lopez-Sole (1995) "Trace elements in coal and their behaviour during combustion in a large power station." *Fuel*, Volume 74, Issue 3, March 1995, Pages 331-343.

ROE GMBH (1982). "Feasibility Study of the Carbonization of Okaba Coal." Final Report prepared for the Nigerian Coal Corporation. Berlin, West Germany.

Udeagbara, S. G. and O. J. Anusiobi (2013). "Quantitative analyses of trace elements in some Nigerian coals." *Australian Jnl. of Basic and Appd. Sc.*, 7(2), pp. 541-46. pp. 541-546.

7 Conversion potentials of Nigerian coals: gasification and briquetting

7.1 INTRODUCTION

The potential of Nigerian coals for power generation and coke production was discussed in depth in the last chapter. All the coals discovered so far are suitable for combustion, steam raising and power generation, but not for metallurgical coking. There is no data available on any systematic investigation of the gasification characteristics of any Nigerian coal. However, based on available data on the technological properties, most of the coals are potentially suitable as feedstock for the production of environment-friendly synthetic gas, liquid fuels, chemicals, smokeless fuel briquettes, and formed-coke. Coal gasification and conversion technologies are treated in depth in another book and only a summary is presented here. (See Afonja, 2017).

7.2 COAL COMBUSTION AND CARBONIZATION

Coal heated in an environment in which air supply is copious will ignite and all the carbonaceous and most of the volatile matter content will burn, yielding energy. The main volatile products of this *combustion* process are carbon dioxide, water vapour and nitrogen oxides, sulphur dioxide and oxygen. The solid residue of coal combustion is ash which is formed mostly from the combustion of the inorganic matter in the coal. However, if air supply is insufficient to ignite the coal, *carbonization* occurs. The chemical reactions involved in the two processes differ significantly, and so is the chemistry of the exhaust products (Table 7.1).

Any coal, both caking and non-caking is potentially suitable for combustion or carbonization. However, a number of variables determine its suitability for an application. Coal required for carbonization and production of metallurgical coke should exhibit a significant amount of thermoplasticity during heating. However, any coal is potentially suitable for carbonization specifically for the production of gas or smokeless fuel, or combustion for energy production. Coals that have some thermoplasticity may cake and present problems in the combustion chamber, hence are often pre-oxidized to remove the plastic properties. Low-grade coals have relatively low carbon content and more coal is required to produce a unit of energy compared with higher grade coals. Also, coals which contain significant amounts of sulphur produce sulphur oxides on combustion or carbonization and expensive cleaning processes may be required to meet environmental requirements.

Table 7.1 Comparison of the primary products of main elemental constituents of coal on carbonization or combustion. Most of the sulfur in the coal is converted to hydrogen sulfide (H_2S) during gasification. However, depending on the gasification temperature and moisture content, approximately 3 to 10% of the sulfur is converted to carbonyl sulfide (COS).

GASIFICATION		ELEMENT		COMBUSTION
CO	$\leftarrow$	C	$\rightarrow$	CO_2
H_2	$\leftarrow$	H	$\rightarrow$	H_2O
N_2	$\leftarrow$	N	$\rightarrow$	NO_X
H_2S	$\leftarrow$	S	$\rightarrow$	SO_X
		O	$\rightarrow$	O_2

7.2.1 Coal conversion to producer and coke-oven gas

Conversion of coal to gas is primarily a carbonization process. However, initial supply of energy is always required to raise the coal bed temperature to 300-350°C and start off the self-sustaining carbon oxidation reactions. The energy supply may be external or generated internally by partial combustion of part of the coal charge. Air is supplied to the reactor but the product gas is of low calorific value due to dilution by nitrogen. Modern gasifiers use mixtures of air and oxygen or pure oxygen, and steam may also be supplied to increase calorific value. The main gaseous products of carbonization are carbon monoxide, hydrogen, nitrogen, some carbon dioxide and hydrogen sulphide, leaving a solid residue of semi-coke or coke, depending on the carbonization temperature. Carbonization up to around 750°C yields semi-coke while full carbonization to 1000°C and over produces coke. The gaseous product at the lower temperature is known as producer or water gas depending on whether air/oxygen and steam are supplied, while full carbonization yields *coke oven gas*.

Coal carbonization yields both gaseous and solid products. From temperatures of around 250-350°C (depending on the coal rank), tar is the main constituent of the gases released but, as temperature increases, the coal becomes incandescent and reactions between carbon, oxygen or hydrogen occur, leading to the production of mainly carbon monoxide but also some hydrocarbons, mainly methane. This first stage of gas release is usually complete around 750°C (*low-temperature carbonization*). However, if the temperature continues to rise, more gases, mainly hydrocarbons and water vapour are released and degasification is complete at 900-1000°C, leaving residual volatile matter of 1-2% (*high-temperature carbonization*).

The gas released from the coal carbonization process up to around 750°C, (producer gas) is of low calorific value (3.5-6 MJ/m³) because it contains about 55% nitrogen from the air. However the heating value can be increased significantly by using oxygen-rich air. Also, the introduction of steam into the input air (*water gas shift reaction*) results in reaction between steam and carbon, with the production of hydrogen and carbon monoxide. The water gas shift reaction is strongly endothermic. When the incandescent coke has been cooled to a temperature at which endothermic reaction can no longer proceed, the steam is replaced with a blast of air. The gas produced is known as water or *blue water gas* (or *syngas* in modern terminology), and has a much higher calorific value than producer gas (11-20 MJ/m³). The solid residue in low-temperature carbonization is known as semi-coke or char.

The residue is soft and may contain up to 22% tar-free, combustible volatile matter. On the contrary, high-temperature carbonization yields hard coke (agglomerated or granulated depending on coal rank and type) with little or no residual volatile matter. Also, the gas produced is of higher calorific value, up to 20 MJ/m³ because of the higher methane content. This gas is primarily a by-product of coke ovens, hence the name, *coke-oven gas*. The typical chemical compositions of the various coal gases are shown in Table 7.2. Coal gasification with the specific objective of producing water or producer gas (low-temperature carbonization) dominated the global energy scene for over a century but has become obsolete as a result of the availability of natural gas, although it is still being practiced in some developing countries. The two most popular types of low temperature coal gasification technologies are the Koppers and Lurgi processes. The two processes were developed specifically for the production of automotive and aviation fuels from coal and advanced versions are at the core of many modern coal-fired power plants all over the world. Some Lurgi plants are still operating in some parts of the world, producing town gas (Figure 7.1).

Table 7.2 Chemistry of gases produced from coal.

COMPOSITION/ HEAT VALUE	PRODUCER GAS	BLUE WATER GAS	COKE OVEN GAS	BLAST FURNACE GAS	SYNGAS	NATURAL GAS
Methane (Vol%)	2	0.8	28-32	< 1	0.1-11.5	70-90
Carbon Monoxide (Vol%)	25	41	5-8	27	12-53	-
Carbon dioxide (Vol%)	5	4.7	2-5	11	15-32	< 1
Hydrogen (Vol%)	15	49	55	2	26-53	0-5
Nitrogen (Vol%)	53	4.5	4	60	0.3-1.5	0-5
Calorific value (MJ/m³)	5-6	10.5-11.5	19-20	3.5-4	15-20	37-43

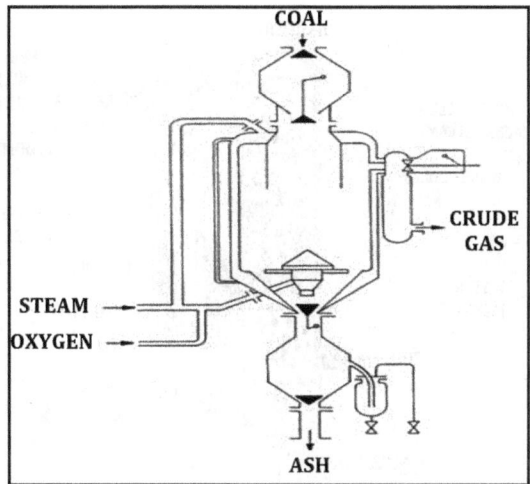

Figure 7.1 A Lurgi coal gasifier.

7.2.2   Coal conversion to synthetic gas (syngas)

Although interest in coal gasification declined from the 1950s, intensive research and development continued in a few countries, notably South Africa. For decades the country has been producing synthetic gas, gasoline, diesel oil, aviation fuel and a wide range of important industrial chemicals from coal. The first coal-to-syngas process known as the Fischer-Tropsch or F-T process was developed in Germany in 1925 as a coal-to-liquid fuel process to convert the country's vast coal resources to liquid fuels. The original gasifier operated at atmospheric pressure but a pressurized version was developed in 1931. The first Lurgi commercial plant was built in 1936 primarily to process lignite and sub-bituminous coals and further development produced gasifiers that could process bituminous coals in the 1950s. Production of coal-based synthetic fuels declined in Germany in the late 1940s but development continued in South Africa under license (SASOL process).

Coal gasification technology is now well developed and several hundred modern plants are operating worldwide. Most are capable of processing other carbonaceous resources such as biomass, oil residue and tar sand. Many gasifiers are operating as part of power generation configuration, others are producing syngas for heating and steam raising, and as feedstock for the production of hydrogen and chemicals, in particular, ammonia and methanol which are precursors to the production of many other valuable chemicals. The modern coal gasifier comprises a closed reaction chamber in which graded coal (6-50mm) is heated in minimum air or oxygen to above 400°C. Gases rich in tar, volatile matter, phenols and gaseous hydrocarbons are released. Some carbon dioxide may also be produced. Considerable heat is released by the carbon-oxygen reaction but the introduction of steam with oxygen/air intermittently promotes a strongly endothermic reaction which produces more hydrogen. Many variants of the coal gasifiers which have been developed can be classified into three main categories: moving-bed, fluidized-bed and entrained-flow bed (Figure 7.2).

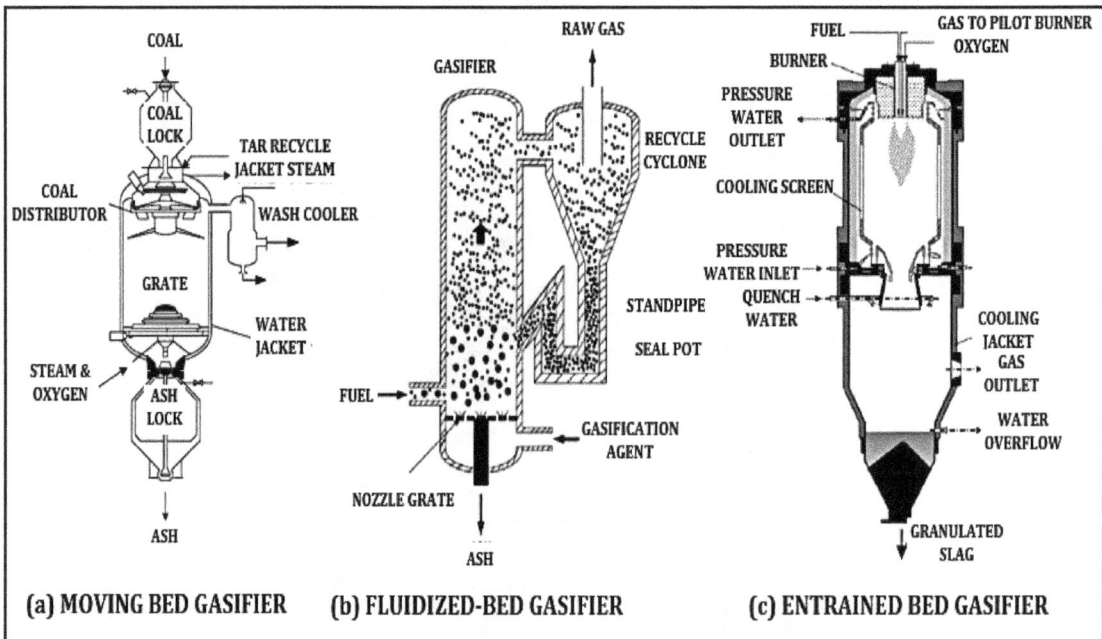

Figure 7.2   Modern coal gasifiers.   *(IEA-Etsap, 2010).*

The moving-bed reactor features a reaction bed of coal moving slowly down the reactor under gravity and a counter-flow of hot blast of air or oxygen, with intermittent introduction of steam. The basic design of the moving-bed gasifier (sometimes called fixed-bed or dry bottom gasifier) has been established by Lurgi since the late 1920s and some of the earliest versions featured pressurized gasification chambers. Most of the development since then has been pioneered by SASOL of South Africa and, since 2001, by Sasol-Lurgi joint venture. One of the most popular versions of the Sasol-Lurgi systems is the fixed bed dry bottom (Sasol-Lurgi FBDB) gasifier which is particularly suitable for processing low-grade coals.

Sasol-Lurgi gasifiers can process high ash coals (up to 50% ROM basis) and the provision of an internal drying zone before gasification zone makes it possible to process high moisture run of mine coals without prior drying. Also, highly reactive low-grade coals can be gasified. This is in fact an advantage since the operating temperature of the gasifier is relatively low (The top and bottom temperatures are around 600°C and 1000°C respectively). The countercurrent flow arrangement enables the heat of reaction from the gasification reactions to be used in pre-heating the coal before it enters the gasification reaction zone and pre-drying of the coal charge is not required. Because oxygen requirement is lower, air is usually used as oxidant and the high nitrogen content can be captured to produce ammonia. This eliminates the need for an expensive air separation unit. Also, the ash bed contributes to agent distribution and provides support within the gasifier (Sasol-Lurgi Technology, 2005).

More advanced variants of the moving bed gasifier operate at temperatures between 800 and 1800°C depending on the coal type and the desired product syngas specifications. The British Gas-Lurgi (BGL) slagging gasifier operates at high pressure and incorporates a molten slag bath at the bottom, hence the common name, slagging gasifier. Syngas production per unit of coal intake is higher compared with a dry bottom gasifier and the quality is higher (IEA-ETSAP, 2010). The temperature in the moving bed reactor is strictly controlled so that the ash does not fuse and drops to the bottom of the reactor as dry ash. Coal with some caking properties can coagulate in the reactor and cause gas flow problems, and may need to be pre-oxidized by drying in air at around 100-105°C. Some systems incorporate stirrers to prevent agglomeration and promote gas flow in the bed.

The fluidized bed gasifier processes ground coal suspended in air on a porous bed Figure 7.2b). This technology promotes excellent heat and mass transfer in the bed due to the intensive mixing. Residence times are shorter than in moving bed gasifiers, hence productivity is higher. However, individual coal particles have varying residence times and a significant amount of unreacted coal is removed from the bed along with the ash of the fully reacted particles. The entrained flow gasifier is similar to the fluidized bed system but gas flow rate is higher and entrained coal particles are captured in cyclones and returned to the gasification zone. Mixing is more intensive and heat and mass transfer more efficient. Most entrained flow gasifiers operate at high pressures of 2 to 7 MPa and temperatures of 1400 to 1600°C. The residence time of a coal particle in an entrained flow gasifier is very small (a few to tens of seconds), hence the gasifier must operate at high temperatures to achieve high carbon conversion. For this reason, entrained gasifiers use oxygen rather than air as oxidant. Entrained flow gasifiers are very often favoured because they can process any type of coal regardless of rank, caking properties or particle size mix. All gasifiers can operate at either atmospheric pressure or at pressures as high as 6 MPa. High pressure systems are preferred in most modern gasifier applications, particularly when included in integrated gasification combined cycle (IGCC) systems because the pressure of the product syngas is sufficiently high to be fed directly to a gas turbine without the need for further compression. Furthermore, capture of carbon dioxide, sulphur and nitrogen oxides, mercury, etc is relatively easy and cheaper compared with other systems. All gasifiers can also operate on air or oxygen as oxidant. While air blown gasifiers are cheaper because of the elimination of the need to install expensive cryogenic air separation units, operating efficiency is significantly lower because the nitrogen in the air acts as a diluent which lowers the calorific value to about a third of that of syngas produced in an oxygen blown gasifier (Philips, 2015; Wang and Stiegel, 2017).

7.2.3   Basic chemistry of coal gasification processes

The chemistry of coal gasification is complex and still a subject of intensive research (Higman, 2008). The major thermodynamically feasible reactions are summarized in Table 7.3. The various reactions can occur and progress to different extents depending on the gasification conditions (coal feedstock, temperature, pressure, oxygen and steam supply). Typically, the supply of oxygen is kept at one-fifth to one-third of the theoretical amount needed for complete combustion of the coal feed. In effect, only partial oxidation of the coal occurs, with the production of carbon monoxide and hydrogen and some carbon dioxide. The carbon monoxide and hydrogen are in turn oxidized (Equations 7.1 a-c, Table 7.3). These reactions are strongly exothermic and the heat released provides the energy required to drive the endothermic reactions (Equations 7.3 and 7.4). Most modern gasification reactors produce a mixture of carbon monoxide, hydrogen and small amounts of carbon dioxide, methane and hydrogen. This synthetic gas or syngas (producer or town gas in older terminology) has a very wide range of applications, from electric power production to liquid fuels, chemicals and high-technology materials. With the addition of sufficient steam in the presence of a suitable catalyst, the carbon monoxide and methane contents could be converted to hydrogen (Equations 7.6 and 7.8).

Table 7.3   Coal gasification reactions. *Afonja, 2017).*

Partial combustion		
$C + O_2 \leftrightarrow CO_2$	$\Delta H = -393.4$ MJ/kmol	[7.1a]
$CO + \frac{1}{2} O_2 \leftrightarrow CO_2$	$\Delta H = -283$ MJ/kmol	[7.1b]
$H_2 + \frac{1}{2} O_2 \leftrightarrow H_2O$	$\Delta H = -242$ MJ/kmol	[7.1c]
Gasification with oxygen		
$C + \frac{1}{2} O_2 \leftrightarrow CO$	$\Delta H = -111.4$ MJ/kmol	[7.2]
Gasification with carbon dioxide [Boudouard}		
$C + CO_2 \leftrightarrow 2CO$	$\Delta H = 170.7$ MJ/kmol	[7.3]
Gasification with steam		
$C + H_2O \leftrightarrow H_2 + CO$	$\Delta H = 130.5$ MJ/kmol	[7.4]
Gasification with hydrogen		
$C + 2H_2 \leftrightarrow CH_4$	$\Delta H = -74.7$ MJ/kmol	[7.5]
Water-gas shift reaction		
$CO + H_2O \leftrightarrow H_2 + CO_2$	$\Delta H = -40.2$ MJ/kmol	[7.6]
Methanation		
$CO + 3H_2 \leftrightarrow CH_4 + H_2O$	$\Delta H = -206$ MJ/kmol	[7.7]
Steam-methane reforming reaction		
$CH_4 + H_2O \leftrightarrow 3H_2 + CO_2$	$\Delta H = 206$ MJ/kmol	[7.8]

Coal, Oxygen, Steam

Further processing to remove carbon dioxide produces pure hydrogen, a valuable fuel for gas-turbine power production and also a major feedstock for the production of chemicals. The synthesis gas produced has a low H_2/CO, around 0.7, compared with the ideal ratio of about 2 or higher required by a number of chemical processes which use syngas as feedstock. These include conversion to liquid fuels by the Fischer-Tropsch coal liquefaction process and conversion into methane, methanol and ammonia. In order to meet the H_2/CO requirement, raw syngas is water-gas shift-reacted with steam (Equation 7.6) often in the presence of a catalyst to increase the hydrogen content. Also, the calorific value of the gas is increased. The syngas has a wide variety of uses including steam raising for electric power production or as feedstock for the production of hydrogen, liquid fuels, and a wide variety of industrial chemicals. The gas produced leaves the gasifier at high temperature and pressure, and is passed through a series of cleaning processes to remove potentially anthropogenic compounds. If oxygen is used in the gasifier instead of air, any carbon dioxide produced is released as a concentrated gas stream in syngas at high pressure because the high nitrogen content of air has been excluded. This makes carbon dioxide capture and sequestration much easier and at lower costs.

7.2.4   In-situ coal conversion to synthetic gas (syngas)

It is widely believed that the global reserves of coal (proven deposits) constitute only a small proportion of the total potential resources. Also, a significant proportion of the reserves is not recoverable by currently available mining technologies because the seams are too deep, too thin, too sporadic, or the geological conditions make recovery by the usual techniques uneconomic or impracticable. Many closed mines still contain considerable coal since only 40-70% of a coal deposit is recoverable by underground mining. Development of in-situ gasification technologies now makes it feasible to convert unmineable and residual coals in closed mines to syngas without the need for expensive mining, transportation and coal preparation processes. Furthermore much of the undesirable anthropogenic particulate and gaseous by-products of coal combustion remain underground.

In-situ underground coal gasification technology (UCG) is not new – early attempts date back to the late nineteenth century. Development was pioneered by the Soviet Union from the late 1920s and the first successful test was conducted in 1934. Producer gas from in-situ coal gasification was in wide local use within a few years. By the 1960s, many commercial plants were running in the Soviet Union but interest declined with the discovery of extensive reserves of oil and natural gas. Development activities in the United States and Europe also declined around the same period for the same reason. The volatility and vulnerability of the oil and gas industry and the political expediency of maximizing internal supply of primary energy have provided the impetus for renewed interest in coal gasification in many countries including the United States, China, Australia, New Zealand and South Africa, and significant progress has been made in the last two decades or so.

The basic technology of in-situ coal gasification involves the drilling of two shafts into a coal seam, one for the injection of oxidant (air, oxygen) and steam into the seam to initiate and sustain the gasification reactions, and the other for the transportation of the gas to the surface (Figure 7.3). The underground reactions take place at very high temperatures from about 700°C to around 1500°C, and high pressures. Unlike a conventional coal gasification reactor, the coal seam is saturated with water and is at hydrostatic pressure which varies

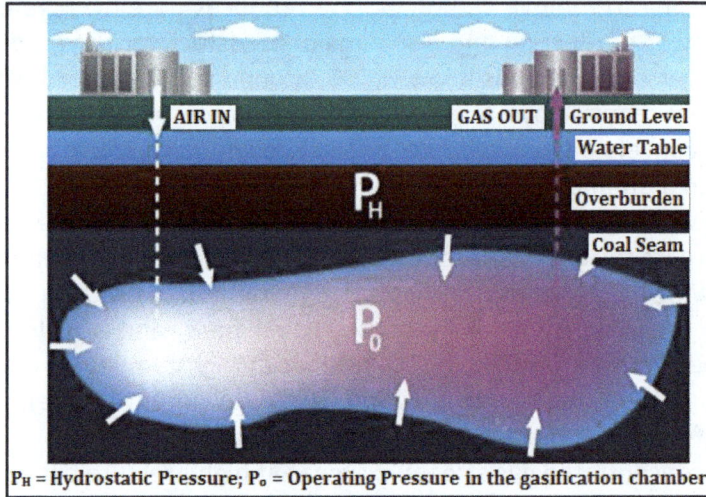

Figure 7.3 The Underground Coal Gasification process. *(www. UCG-GTL.com).*

with depth of the coal seam. A typical UCG operates at slightly below hydrostatic pressure but the pressure could be increased by artificial pressurization to improve gasification rate. The syngas produced underground may be transported by pipeline to consumers or utilized on site for power production. The current status of technology and deployment of UCG has been reviewed extensively in recent literature (Bell et al, 2011; Brown, 2012; Bhutto et al., 2013). Coal seams deeper than about 600 metres previously considered unmineable, seams considered too thin or too discontinuous, and low-grade coals which are deposited too deep underground and cannot be recovered economically by mining, can now be gasified in-situ. Estimates of the economic implications of in-situ coal gasification vary widely, but there seems to be a general consensus that proliferation of the technology could increase the world's recoverable reserves of coal by 200-300%.

7.2.5   Conversion of coal-bed methane [CBM] to synthetic gas (syngas)

Methane (CH_4) is a naturally occurring gas produced in a variety of ways. The coalification process is perhaps one of the major sources, with the gas being formed during the partial decay of organic matter at the peatification stage. Industrial processes and a number of agricultural processes such as ranching and rice farming produce methane. The gas is also being produced for use as domestic gas by fermentation of biowaste. Methane is the major component of natural gas (around 95%). Coal bed methane is natural gas trapped in coal beds. Large quantities of the gas are generated during coalification and stored within the coal on internal surfaces. It is estimated that, because of the large internal surface area of coal deposits, the storage capacity of methane per unit rock volume in coal beds is around seven times as large as a conventional natural gas reservoir. Many coal deposits have methane gas trapped in the seams and cracks, and coal surfaces, held in place by water pressure. The gas is often released during mining operations and is a major potential cause of explosion in coal mines. When the coal deposit lies at shallow depths coalbed methane can be recovered by drilling inexpensive wells.

However, at greater depths, increased water pressure closes the cracks in the coal bed and reduces permeability and the ability of the gas to move through the cracks and out of the coal bed.

Although the extent of coalbed methane as a potential energy resource is believed to be massive and widely distributed globally, scientific understanding of, and production experience with coal-bed methane are limited and intensive research is ongoing. A lot still needs to be learned about the occurrence and recoverability of the resource, and the geologic, geochemical, engineering, technological and economic factors which are significant in the production and utilization of coalbed methane. The environmental implications of coal-bed methane are also not clear. Coalbed methane has been the cause of many major mining accidents but many mines now have vents which exhaust the gas to the surface and to the atmosphere where it may remain for around 15 years, reinforcing the greenhouse effect. More recently, its value as a potential source of primary energy has been realized and intensive research is ongoing on utilization technologies.

Coalbed methane can be recovered from underground coal before, during or after mining operations and various types of technologies are commercially available. The gas can also be extracted from coal seams which are considered unmineable because the seams are too deep, too thin, or of poor or inconsistent quality. The most common method of recovering coal bed methane from coal seams is to pump out the water from the mine. The drop in reservoir pressure releases methane which is pumped to the surface with the water where it is separated from water and transported through pipelines to storage facilities or shipped (Figure 7.4).

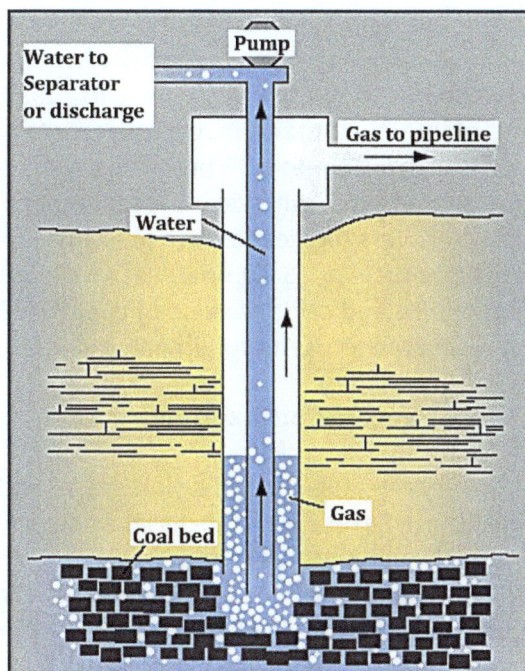

Figure 7.4 Coal bed methane recovery.
(carleton.edu/research) _education/cretaceous/coalbed.html)

Another production technology converts coal-bed methane to a mixture of carbon monoxide and hydrogen (syngas) by reacting the gas with steam in-situ to obtain hydrogen and carbon monoxide. The chemical reaction is shown in Equation 7.9 which is a reverse of the methanation process (Equation 7.7)

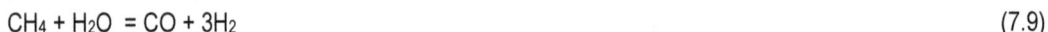

$$CH_4 + H_2O = CO + 3H_2 \tag{7.9}$$

7.2.6   Coal conversion to synthetic natural gas (SNG)

Synthetic natural gas may be produced from syngas by a process of methanation which converts the carbon oxides and hydrogen in syngas to methane by the reactions in Equations 7.10a and 7.10b.

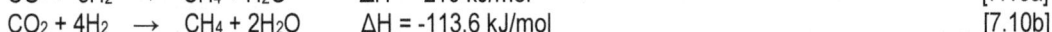

$$CO + 3H_2 \rightarrow CH_4 + H_2O \qquad \Delta H = -210 \text{ kJ/mol} \tag{7.10a}$$
$$CO_2 + 4H_2 \rightarrow CH_4 + 2H_2O \qquad \Delta H = -113.6 \text{ kJ/mol} \tag{7.10b}$$

The reactions take place over nickel catalysts in fixed-bed reactors and are strongly exothermic hence the heat of reaction must be strictly controlled and a catalyst that remains active after prolonged exposure to high temperature is required to avoid catalyst sintering and decomposition of the product methane to carbon. Also, prior to methanation, any sulphur in the syngas must be removed to avoid catalyst poisoning. Coal conversion to synthetic natural gas technology is highly attractive from the perspectives of the environment, energy security and use of available local resources. Natural gas is only available in relatively few countries whereas many countries have significant coal resources. Furthermore, any grade of coal can be converted to environment-friendly SNG. Coal-to-SNG technology is well established and commercial plants have been in operation since the early 1980s. One of the key challenges has been the design of reactors that are capable controlling the adiabatic reaction temperature rise which is high in coal-to-SNG processes. Most of the reactors in commercial operation are of the fixed-bed type, with several reactors operating in series. A design by Topsoe (TREMP process) recovers the heat as high pressure superheated steam (Figure 7.5). The TREMP technology produces SNG compatible with natural gas specifications, with methane content of up to 98% and calorific value of over 37-39MJ/m³. Also, around 85% of the heat of reaction is recovered as useful superheated steam (Jensen et al., 2011).

7.2.7   Coal-based syngas and SNG utilization

Coal gasification plant is very versatile and can be integrated with power, and liquid fuels and chemical production plants in a flexible way. One gasification plant can produce gas for several parallel production plants (Figure 7.6). Apart from use for power generation clean syngas can be processed in a variety of ways to obtain hydrogen, synthetic natural gas, different types of chemicals, liquid fuels and valuable industrial materials (Figure 7.7). For example, clean syngas can be hydrogenated in an F-T reactor at different temperatures and over different catalysts to obtain a wide variety of products, mainly diesel oil (Figure 7.8). A variant of the F-T process (DME process) produces methanol from syngas and further processing produces gasoline, propane and a wide variety of feedstocks for the production of chemicals.

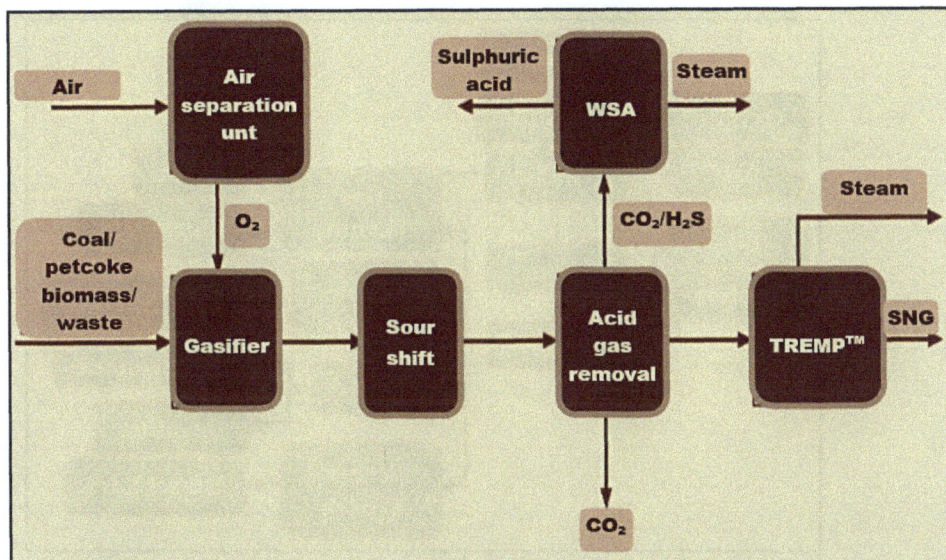

Figure 7.5 TREMP Coal-to-synthetic natural gas flow chart. *(www.topscoe.com).*

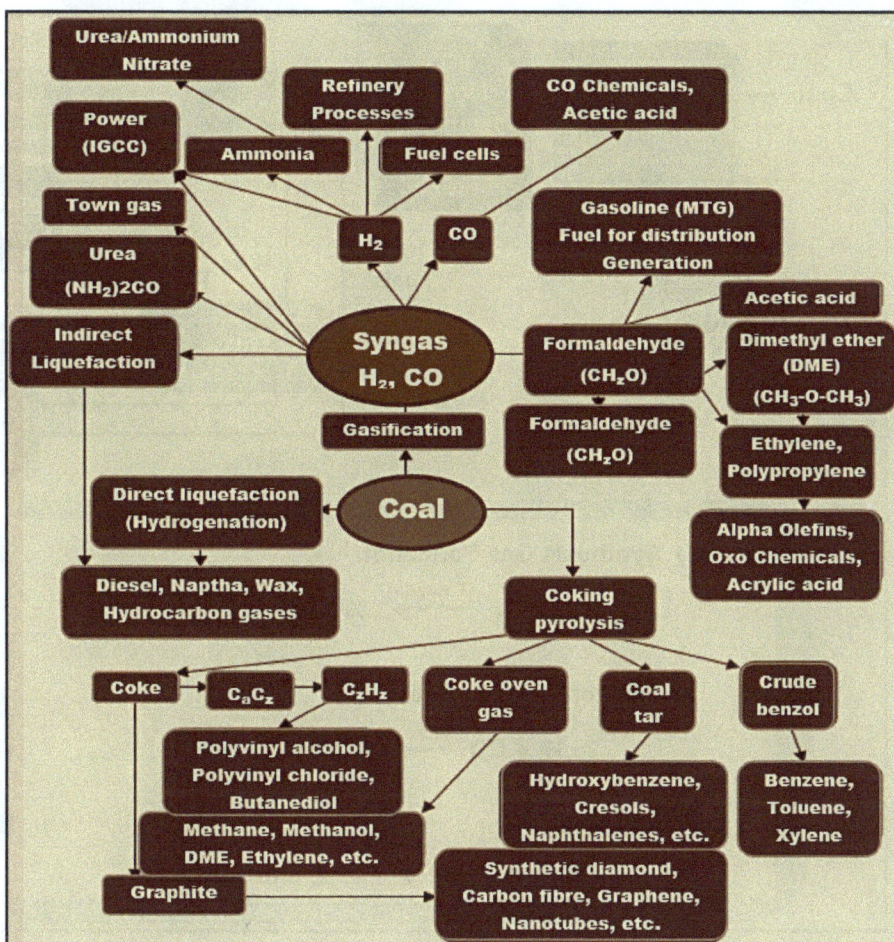

Figure 7.6 Options for coal conversion to energy, premium chemicals and materials.

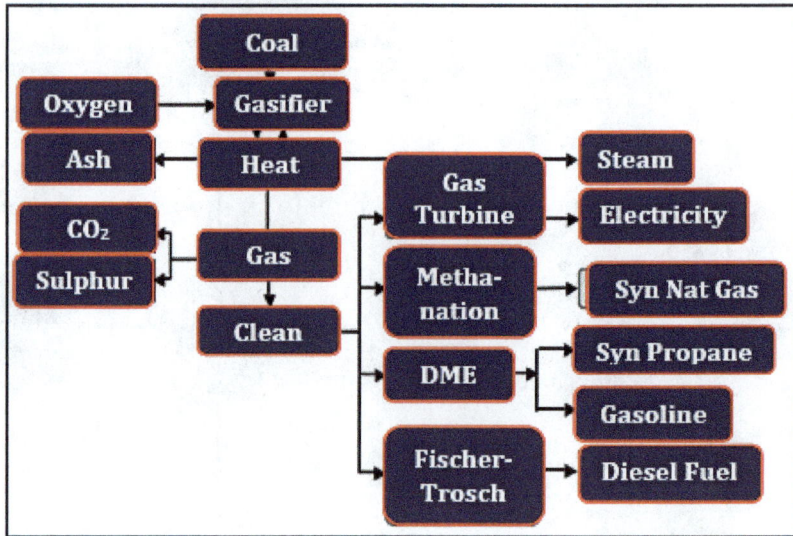

Figure 7.7 Schematics of a typical flexible coal-based syngas production plant.

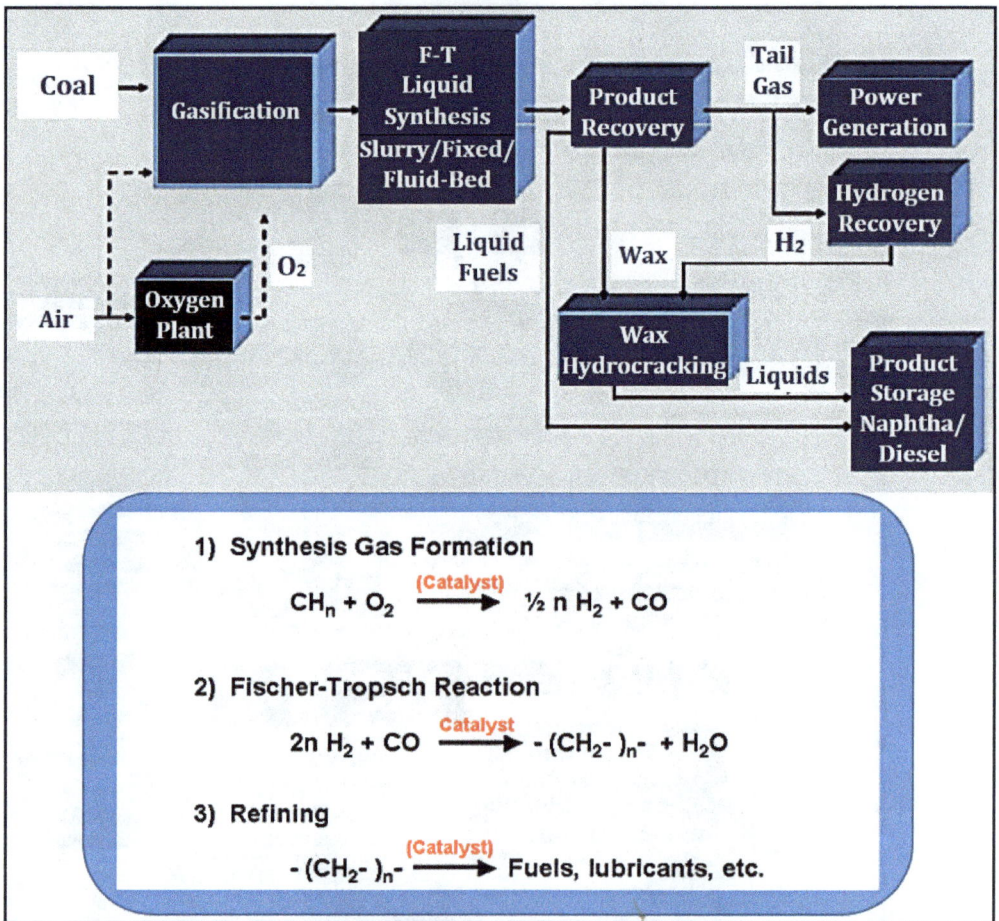

1) **Synthesis Gas Formation**

$$CH_n + O_2 \xrightarrow{\text{(Catalyst)}} \tfrac{1}{2}\,n\,H_2 + CO$$

2) **Fischer-Tropsch Reaction**

$$2n\,H_2 + CO \xrightarrow{\text{Catalyst}} -(CH_2\text{-})_n\text{-} + H_2O$$

3) **Refining**

$$-(CH_2\text{-})_n\text{-} \xrightarrow{\text{(Catalyst)}} \text{Fuels, lubricants, etc.}$$

Figure 7.8 The Fischer-Tropsch coal-to-liquids process.

In spite of the versatility of syngas, it is used mostly in the vicinity of generation because it is not economical to transport the gas through pipelines over long distances due to the low calorific value compared with natural gas. Methanation converts the hydrogen and carbon monoxide contents of syngas to methane, raising the chemical composition and calorific value to a level that meets typical natural gas specifications. Conversion of syngas to synthetic natural gas (SNG) is becoming increasingly widespread, with China leading the world. China has very low natural gas deposits (natural gas contributes only about 4% to the country's primary energy requirements). However, the country hosts one of the largest coal resources in the world, located mostly in the remote northern regions, hence the coal has to be transported at great expense to the industrial southern regions where it is needed. This makes conversion to syngas and SNG very attractive.

Syngas can be produced in the coal regions and utilized for power generation which can be fed to the national grid, or converted to high-value chemicals for internal use or export. For example, the country supplies about a quarter of the global requirements of methanol produced from coal. Synthetic natural gas can also be transported by pipeline over very long distances to the industrial areas. Another advantage of coal-to-gas is that the relatively low pollution products (compared with coal combustion) is confined to a relatively small area of the country and can be controlled more effectively and economically. Conversion of coal to syngas is considered one of the most flexible technologies to produce clean burning fuel for power generation and many modern power plants now feature   coal-to-syngas units. Syngas also has high potential as feedstock for the production of clean hydrogen, liquid fuels and a wide variety of chemicals.

Many of the processes that utilize syngas (in particular, production of hydrogen, liquid fuels and chemicals) require the removal of sulphur content to avoid catalyst poisoning. In coal combustion, sulphur and nitrogen contents are converted to sulphur oxides and nitrogen oxides respectively. In contrast, the products of carbonization are hydrogen sulphide (H_2S) and gaseous nitrogen (N_2) respectively. Some carbonyl sulphide (COS) may also be produced. These products and other undesirable constituents of syngas (particulate matter, trace elements associated with both organic and inorganic components in the feed, such as mercury, arsenic and other heavy metals) can be removed from syngas with relative ease and greater efficiency prior to combustion. In coal-fired plants, removal of these compounds after combustion is much more difficult and expensive.

Another major asset of coal-to-gas technology is the ability to process any type of coal in particular, low-grade coals which are of relatively low commercial value. Furthermore, a syngas plant can be an integral unit of a power generating plant and over a thousand coal-based syngas plants are in commercial operation in different parts of the world.

7.3   CONVERSION OF NIGERIAN COALS TO SYNGAS

Gasification of non-caking coals and other coals that have low market value is becoming increasingly widespread because deposits are abundant in many countries and the added cost of gasification can be easily absorbed. Coal produced cheaply by opencast mining is particularly suitable and high ash coals (20-50% wt), low-rank coals (brown coals and lignite) are of special interest for gasification. Low grade sub-bituminous coals and lignites have also been found very suitable for underground gasification because of the relatively high seam porosity which makes it easy to link injection and production wells.

Most Nigerian coals are low-rank and non-caking and, although they are suitable for local utilization in coal-fired power plants, the export potential is not very high. One of the main parameters that determine the export value of coal is the cost of transportation per unit heat content. Because low-grade coals have high density, high water content, low carbon content and low heat content per unit weight, this ratio is very high. However, the potential for gasification is excellent. Most of the coals have very low sulphur content and several of the deposits can be exploited by opencast mining.

In view of current efforts to phase coal back into the energy mix and the country's commitment to the global clean energy movement, integration of coal gasification should be made one of the conditions for approval of future coal-fired generating plants. In any case most modern coal-based power plants now feature coal gasification as the first stage and all the power plants currently at the planning stage will include coal gasification. Adoption of coal gasification for future coal power plants could also open up a vast range of new potentials for Nigerian coals since the gasification unit of a power plant can be designed to be flexible so that the syngas can also be used as feedstock for a wide range of other applications as shown in Figure 7.6.

Nigeria has vast resources of natural gas, considered to be more than crude oil resources, hence conversion of coal to synthetic natural gas (SNG) may not be an attractive option. However, in-situ coal gasification for power generation could be a cheaper option to underground mining and expensive above ground coal preparation plant would not be required, but the geological and geophysical structure of each deposit would need to be evaluated in order to determine the suitability for in-situ gasification.

7.4   CONVERSION OF NIGERIAN COALS TO SMOKELESS FUEL

Coal can be thermally processed to remove some or all the volatile matter, either to produce smokeless fuel or formed-coke for metallurgical applications. Other uses of devolatilized coal include production of activated carbon and carbon sieves. Coal is not considered a clean fuel because of the objectionable gaseous and environmentally harmful emissions. Many processes have been developed over the last hundred years or so to 'desmoke' coal by low-temperature carbonization followed by briquetting to produce clean domestic and industrial fuel. Partial devolatilization removes objectionable tar but retains the combustible volatile matter content. The solid product of coal carbonized at low-temperature retains much of its reactivity and smoke-free volatile matter content, it is readily lightable and combustible, making it suitable for the production of smokeless fuel. When carbonized at high temperature, nearly all the volatile matter is expelled or burned, with residual volatile matter less than 2%. The solid product is suitable for the production of formed-coke for metallurgical applications.

7.4.1   Basic smokeless coal production process

Coal may be converted to smokeless fuel by controlled thermal treatment of granules of the coal up to about 550°C, a process which facilitates the release of tarry gases and retention of combustible gases, the primary objective being to convert a low-grade coal into a fuel product of superior quality. The residual char may be in lump form if the original coal has caking properties, or remain as granules if the coal is non-caking. The product of low temperature coal carbonization known as char or semi-coke is a valuable fuel and can be used to fire boilers or fuel other industrial or domestic heating appliances. The bulk of coals converted to smokeless fuel worldwide is of the non-caking type (low grade coal or anthracite) and the char is often agglomerated by compaction under pressure (briquetting) with or without the addition of a binder. However, in most cases the char is mixed with a binder such as pitch, starch, molasses, etc and formed into briquettes in a press. Several binderless briquetting processes were patented in the 1930s, all involving heating finely divided coals to an appropriate temperature and hot briquetting directly on discharge. The processes were based on the theory that every coal passes through a plastic stage on heating, usually between 300 and 450°C. The softening process also coincides with the release of tarry volatile matter which acts as binder during hot briquetting in the plastic temperature range. Some smokeless fuel production processes involve pre-compaction prior to carbonization while others, in particular those which adopt fluidized bed carbonizers are compacted after carbonization using binders and the product briquettes may be given secondary heat treatment to improve strength.

The briquetting process has been used for over a century for producing compact, high-calorific value fuel from low-grade friable coals for domestic and industrial use. The process is not confined to coals alone and is applied to various powders, pharmaceutical products, ores, and agricultural waste products. More recently, the coal briquetting process has been adopted for the production of metallurgical quality coke from low-grade coals. Also, coal fines which otherwise would have gone to waste and friable coals are often agglomerated into more valuable fuel. Most commercial processes involve grinding raw coal or semicoke, sizing and drying the coal, mixing with a binder which may be coal tar pitch, bitumen, caking coal, lime or starch, heating the mixture and compacting in presses into solids of suitable shapes and sizes, and cooling. Both the pre- and post-compaction carbonization may be carried out in retorts or fluidized-beds and the briquettes produced may be further heat-treated to improve strength. Most commercial processes feature double roll presses of the type shown in Figure 7.9 which produce pillow-shape briquettes. Critical process variables include the physical properties of the coal, its surface chemistry, and the thermochemical properties of the coal. The devolatilization behaviour in the temperature range 300–550°C is crucial since the primary objective is to get rid of the tarry gaseous substances while the clean combustible gases such as hydrogen, carbon monoxide and methane are retained, otherwise the briquettes would have poor lightability.

BRIQUETTING MACHINE   **BRIQUETTING ROLLS**

Figure 7.9 A Komarek high-pressure briquetting and compacting machine. *(Komarek.com).*

7.4.2 Smokeless fuel production from Nigerian coals

Extensive investigations have been carried out to determine the carbonization and briquetting characteristics of Nigerian coals (Arthur D. Little, 1965; Afonja, 1972, 1973, 1977, 1979, 1981a, 1981b, 1983, 1986a, 1986b, Olofinjana and Afonja, 1988, 1996). The tests focused on the feasibility of producing briquettes as smokeless fuel or blend components in metallurgical coking. The feasibility of incorporating Nigerian coals as coking blend components has been discussed exhaustively in Chapter 6 and the focus here is on the feasibility of producing smokeless fuel and metallurgical formed-coke from the coals. In 1965, the Federal Ministry of Industries commissioned a study on the feasibility of large-scale carbonization of Enugu coal. The primary objective was to revitalize the coal industry by diversifying the products and expanding the market potential for the products of carbonization (smokeless fuel, tar, gas, chemicals) to most parts of Nigeria and other West African countries. The report submitted by Arthur D. Little (1965) was based on a theoretical projection from results of laboratory tests and recommended the adoption of large-scale carbonization to produce smokeless fuel and chemicals. However, no further action was taken on the report.

Further to the Arthur D Little study, extensive tests have been carried out on Enugu, Okaba and Obi/Lafia coals to determine the feasibility of production of smokeless fuel (Afonja, 1974; 1979; 1983; Olofinjana and Afonja, 1988). One major requirement for good smokeless fuel production is that char of desired quality must be produced. The process of devolatilization of coal must be controlled such that the volatile tar component is expelled while sufficient combustible volatile matter must be retained to enhance lightability. Tests were carried out on a laboratory fluidized bed described in Chapter 6 (Figure 6.3). Chars were prepared at various temperatures between 350°C and 550°C and different residence times. The residual volatile matter content of each treated sample was determined and the residues were chloroform-extracted to determine the optimum conditions for obtaining the desired residual plasticity. (Figures 7.10 and 7.11). The chars were mixed with pitch or starch binders at varying proportions and formed into briquettes Chars prepared at various temperatures were mixed with pitch of starch binders at varying proportions and formed into briquettes at different roll pressures on a Komarek T100 pilot plant briquetting press (Figure 7.9). Briquetting tests were also carried out on binderless chars in heated cylindrical moulds at different temperatures and pressures on a laboratory Carver press. Chars made from all the coals by cold briquetting with 8-10% soft pitch or about 15% starch produced strong briquettes but those with starch binders deteriorated in strength within a few days, due presumably to hydration, hence, all subsequent tests were carried out using pitch binder. The briquettes were carbonized in a box oven purged with nitrogen for 30 minutes at varying temperatures and the residual volatile matter at each temperature was determined.

The strength of the briquettes formed on the roll press was determined by the standard shatter test which involved dropping 5kg of briquettes four times through 2 metres on a steel plate and measuring the weight percentages of +300mm briquettes. The binderless briquettes formed in cylindrical moulds were tested for strength on the Carver press. Lightability and smoke production tests were also performed on the briquettes produced from Nigerian coals under various conditions. Air was blown over a small heap of briquettes piled over a lit kerosene-soaked wool. The time it took for the briquettes to become incandescent was taken as a measure for comparing lightability of briquettes made under different conditions. The colour of the smoke was determined using a photoelectric cell to observe the colour of a beam

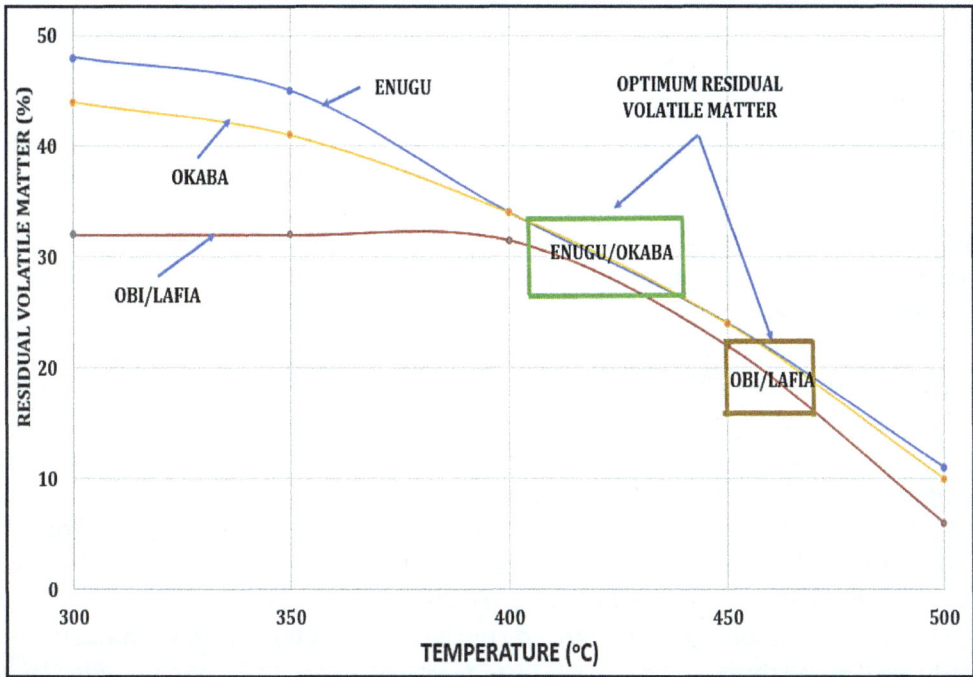

Figure 7.10 Effect of fluidization temperature on devolatilization of Enugu coal. *(Afonja, 1974, 1979, 1983).*

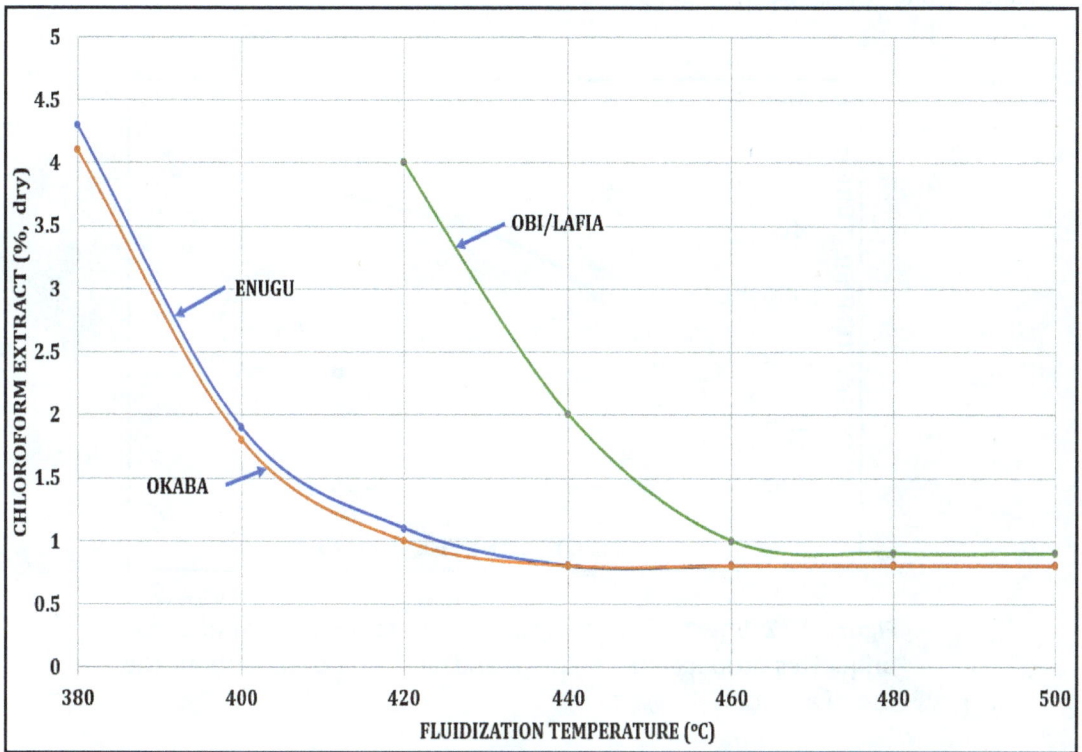

Figure 7.11 Effect of fluidization temperature on the chloroform extract yield of Enugu and Obi/Lafia coals. *(Afonja, 1979).*

of light traversing the smoke emission from the burner. Difficult lighting was indicative of low residual volatile matter while blue smoke indicated the presence of residual tar.

Fluidization temperature and residence time were found to be the major determinants in producing chars that yielded strong smokeless briquettes with good lightability, irrespective of whether or not a binder was used. The optimum parameters established for Enugu blend component char and presented in the last chapter were a carbonization temperature in the range 410 to 440°C, which corresponded to a residual volatile matter in the range 34 to 25% and an optimum residual plasticity characterized by a chloroform extract in the range 1.2–0.8%. Optimal residence time at the peak fluidization temperature was 15-20 minutes. Similar results were obtained for the same coal in the current tests and there was no significant difference for Okaba coal. However, Obi/Lafia coal had very different devolatilization characteristics (Figure 7.11), the optimum temperature range being 450–465°C corresponding to a residual volatile matter content of 15-22%. The results of shatter tests on the binderless briquettes are presented in Figure 7.12. The strongest briquettes were produced if briquetting was carried out at 400–450°C and there was no significant difference between the three coals. Post compaction heat treatment improved briquette strength but the optimum range for good lightability was also in the briquetting temperature range (Figure 7.13). The products are shown in Figure 7.14.

The results presented above proved conclusively that good quality smokeless fuel can be produced from all three Nigerian coals tested, with or without a binder, provided that the chars are produced under optimum conditions. However, briquetting on a large scale without a binder presents operational problems because the briquetting has to be carried out at elevated temperatures. All the coals contain high volatile matter that yields considerable tar from which suitable pitch binders can be produced.

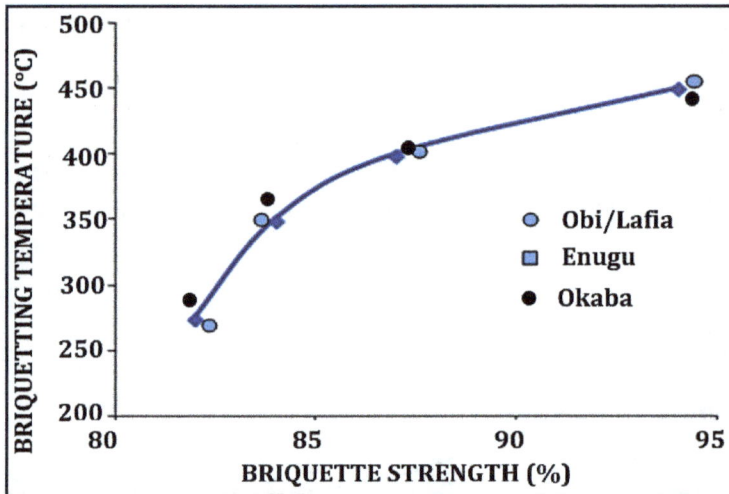

Figure 7.12 Effect of temperature on the strength of smokeless briquettes produced from Enugu and Obi/Lafia coals without the use of a binder. *(Afonja, 1979).*

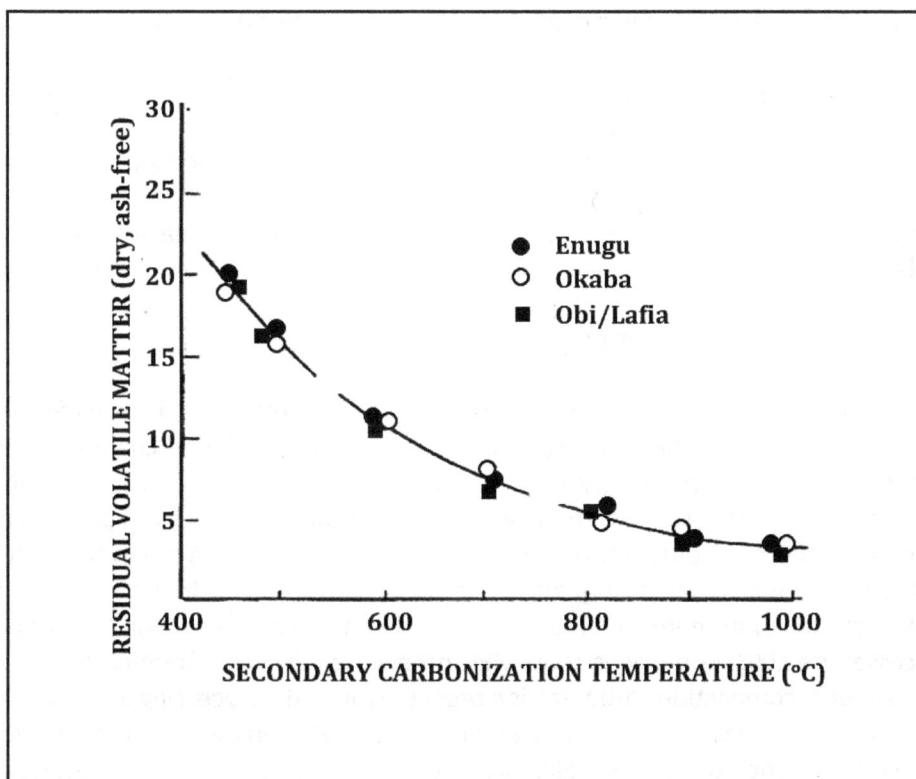

Figure 7.13   Effect of secondary carbonization temperature on the residual volatile matter of briquettes made by binderless briquetting of Obi/Lafia, Enugu and Okaba coals. *(Afonja, 1979).*

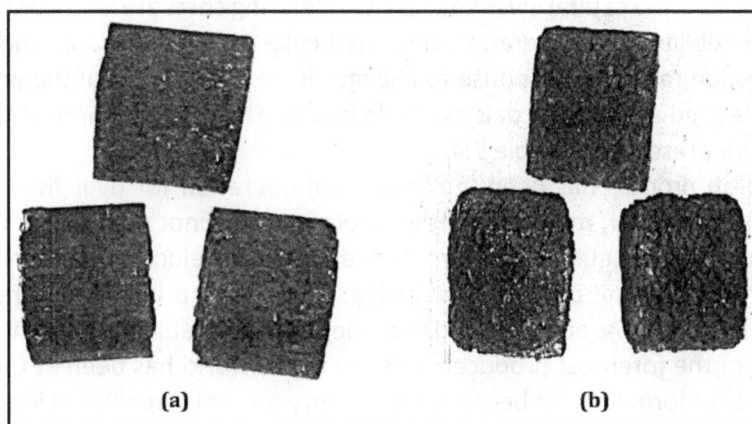

Figure 7.14   Briquettes from Enugu coal. (a) green briquettes; (b) calcined briquettes. *(Afonja, 1974; Olofinjana and Afonja, 1988).*

7.5   CONVERSION OF NIGERIAN COALS TO formed-coke (OR FORMCOKE)

Coking coal is the most expensive raw material for iron production by the blast furnace route and intensive research has focused on optimizing its use by supplementing with less expensive fuels such as coke breeze, briquetted non-coking coals, coke oven gas or natural gas. Efforts have been made also to develop new, non-conventional processes for the production of coke of metallurgical quality (formed-coke) from globally abundant weakly-caking and non-caking coals. Many processes have been patented and some have reached commercial stage.

7.5.1   Formed-coke technologies

Many technologies for conversion of low grade coals to metallurgical quality formed-coke have been developed over the last fifty years or so. Basically, crushed coal is carbonized at low temperature to obtain char (semicoke) which is mixed with a binder and compacted into briquettes. The briquettes are then heat treated to obtain the desired physical and chemical properties. The main differences between the various processes are in the coal type or mix, method of compaction (hot briquetting, cold briquetting, pelletizing), the binders used (coking coal, pitch, lime, bitumen), and the nature of pre- and post-compaction heat treatment. Some processes use 100% non-coking coals while others use mixtures of non-caking and caking coals. Pre- and post compaction carbonization processes include processing in fluidized-beds, rotary kilns, multi-stage flash heaters, vertical retorts, vertical ovens, and sand beds. Most processes involve compaction of the semicoke mixed with proprietary binders in briquetting presses.

Formed-coke has a number of potential advantages over conventional coking: it is capable of utilizing a wide range of coals including non-coking coals; the size and shape of the briquettes can be varied to achieve optimum bed permeability in the blast furnace; formed-coke of consistent quality can be produced; the potential for continuous, automated production is very high; coke yield is higher than in conventional coking. Atmospheric pollution in formcoking is considerably less and capital investments and operating costs are less. Another feature of great interest is flexibility in the operation of formed-coke processes, that is, the ability to start and stop production rapidly in response to changes in demand. Most of the processes developed so far have aimed at utilizing coals available locally. The main features of the most prominent processes are presented in Table 7.4.

The Polish process has been in commercial operation for over five decades, producing foundry coke. However, most of the other processes have not reached continuous commercial production mainly because blast furnace operators are reluctant to try new materials which may require a radical departure from well established and familiar operating conditions for their blast furnaces, especially as most of them have reliable sources of conventional coke. Japan, one of the foremost producers of steel in the world has been at the forefront of blast furnace trials of formed-coke because the country has only weakly-caking coals and is heavily dependent on imported coking coal. Japan imports approximately 90 types of coals from six countries and 10 to 15 different coals are blended for typical coke oven charges.

In view of the increasing scarcity of prime coking coals and the continued dominance of the iron blast furnace technology, it is inevitable that formed-coke will in future become an important source of metallurgical coke. Table 7.5 presents a comparison between conventional oven coke and formed-coke.

Table 7.4 Main formed-coke processes.

PROCESS	MAIN FEATURES
SAPOZNIKOV (RUSSIA)	Weakly-caking coals flash-heated in a multi-stage heater and briquetted at 400-500°C without the addition of a binder. Briquettes are carbonized at 850°C and gas-cooled
FMC (USA)	Weakly or non-caking and sub-bituminous coals oxidized and carbonized at 850°C in multi-stage fluidized-beds and briiquetted with 10-15% pitch. No binder is required if the volatile matter of the coal is greater than 35%. Briquettes are oxidized at 250°C, carbonized at 850°C and gas-cooled
ICEM (ROMANIA)	Weakly and non-caking high volatile coal briquetted, using medium-caking gas coal and pitch as binder. Briquettes carbonized at 950 – 1000°C in shaft oven.
BFL (GERMANY)	high-volatile non-caking coal carbonized in fluidized-bed at 700°c and briquetted at 400 – 500°c using caking coal and recycled by-product as binder.
AUSCOKE (AUSTRALIA)	Weakly-caking low and medium-volatile coal mixed with pitch binder and cold-formed. Briquettes carbonized at 500°C in fluidized-sand bed and subsequently coked in directly-heated shaft oven at 1100°C.
DKS (JAPAN)	Low-volatile weakly caking coal mixed with 5 to 10% coking coal, 0 to 10% coke breeze and 10% binder consisting of pitch and tar, and cold-formed in a roll press. Briquettes are coke in an externally-heated shaft furnace at 1250 to 1300°C for 10 hours and water-quenched.
HBNPC (FRANCE)	Wide range of coals including non and strongly caking coals mixed with pitch binder, cold-formed and carbonized to 1000°C in a continuously or semi-continuously operated vertical oven with internal heating.
I.Ch.P.W. (POLAND)	Lump coal carbonized at 750 to 800°C, crushed and briquetted with 10% binder consisting of 70% pitch and 30% heavy oil. Briquettes oxidized at 200 to 300°C and water-quenched.

Table 7.5 Comparison of the properties of oven coke and formed-coke. *(compiled from Afonja, 1979, Aderibigbe and Szekely, 1982).*

PROPERTY	OVEN COKE	FORMED-COKE
Crushing strength (kN/m^2)	6,895	12,000
Residual volatile matter (%, daf)	1.5-2	5-10
Density (g/cm^3)	1.1	1.42
Specific surface area (m^3/g)	1.3-1.6	40
Porosity (%)	51-56	35-45
Average pore diameter (μm)	9.6-10.8	2.07

Generally, formcoking allows better control of the final properties of fuel than conventional coking. It is possible to produce formed-coke of comparable quality to beehive oven coke but, in general, formed-coke is stronger, contains more volatile matter and is more reactive. The higher reactivity of formed-coke is thought to be due to the extent of graphitization that has been attained during production rather than the effect of specific surface (Aderibigbe and Szekely, 1982). It is believed that increased reactivity due to higher specific surface is compensated for by loss in reactivity due to lower porosity or average pore diameter. This then suggests that formed-coke of comparable reactivity can be produced if more time is allowed for graphitization during production. The higher reactivity of formed-coke is however not a disadvantage as regards its use as a blast furnace fuel.

Formed-coke has been tested extensively in commercial blast furnaces though for relatively limited durations due to insufficient availability of the product (Ahland et al, 1971; Nekrasov, 1974; Dartnel, 1978; Kohno, 1988). The Japan Iron and Steel Federation (JISF) process shown in Figure 7.15 is a 200t/day facility which adopts a continuous process in which mixtures of non-caking and weakly caking coals are briquetted and carbonized in a gas heated vertical shaft furnace, cooled in the lower part of the furnace and discharged at the base. The products of the JISF process were tested continuously in a large 4,250m³ commercial blast furnace for 110 days (Kohno, 1988). The results of the tests proved that formed-coke produced entirely from non and weakly caking coals can be used in commercial blast furnaces without any noticeable deterioration in performance indices.

A new formed-coke technology whose development began in the late 1980s has reached an advanced stage and commercial tests are in progress. NuEnergy process produces metallurgical quality formed-coke (NuCoke) from a wide variety of coals (Cordova et al., 2014). A 10-tonne pilot plant has been operating in Virginia, U.S.A since the early 1990s and another one was built in Arkansas, USA in 2009. Products of the pilot plants have been tested extensively in commercial operations over several years and a 50,000-tonne commercial plant is under construction. The first stage of the NuEnergy process involves char production from ground, dried coal carbonized at low heating rate between 427°C – 650°C, for 20 minutes. The char is stored in bins and the byproduct gas is processed to recover valuable chemicals and heat. Char yield varies depending on the feed coal. Low volatile bituminous coal produces over 80% char compared with around 70% for high-volatile bituminous thermal coal and 35% for sub-bituminous coal. On the other hand sub-bituminous coal yields more byproduct gas. Clearly, high volatile coals are the most suitable for the process in terms of cost and char yield, since low-volatile coals are expensive and scarce.

In the second stage, the char is mixed with coking coal and tar pitch binder in the ratio 60:30:10 and compacted in a briquetting plant. The briquettes are calcined in a continuous oven with several heating zones at around 1100°C. The NuCoke produced is cooled in a quenching tower. A schematic flow diagram is shown in Figure 7.16. The char (NuCarbon) produced in the NuCoke process has superior properties compared with anthracite which it can replace in many applications. Being a closed and continuous process with internal incinerator, demister, and scrubber units, gaseous emissions and particulate matter are minimized and potentially anthropogenic metals such as mercury are removed.

The results of testing the JISF formed-coke in a commercial blast furnace showed that formed-coke was about 15% cheaper in both facility and manufacturing costs than the conventional coking process. Furthermore, because the process is continuous, it is more amenable to automatic control and is superior to conventional coking technology in terms of utilization of non-coking coals, manpower savings, and environmental degradation. The NuEnergy product compares favourably with conventional blast furnace coke (Table 7.6). The product has also been tested extensively in several iron and steel plants and foundries in the USA and performed satisfactorily. NuCoke also has great potential as carburizer in steel ladles, as fuel in direct reduction iron and steel process, and as low-pollution fuel for electric power generation.

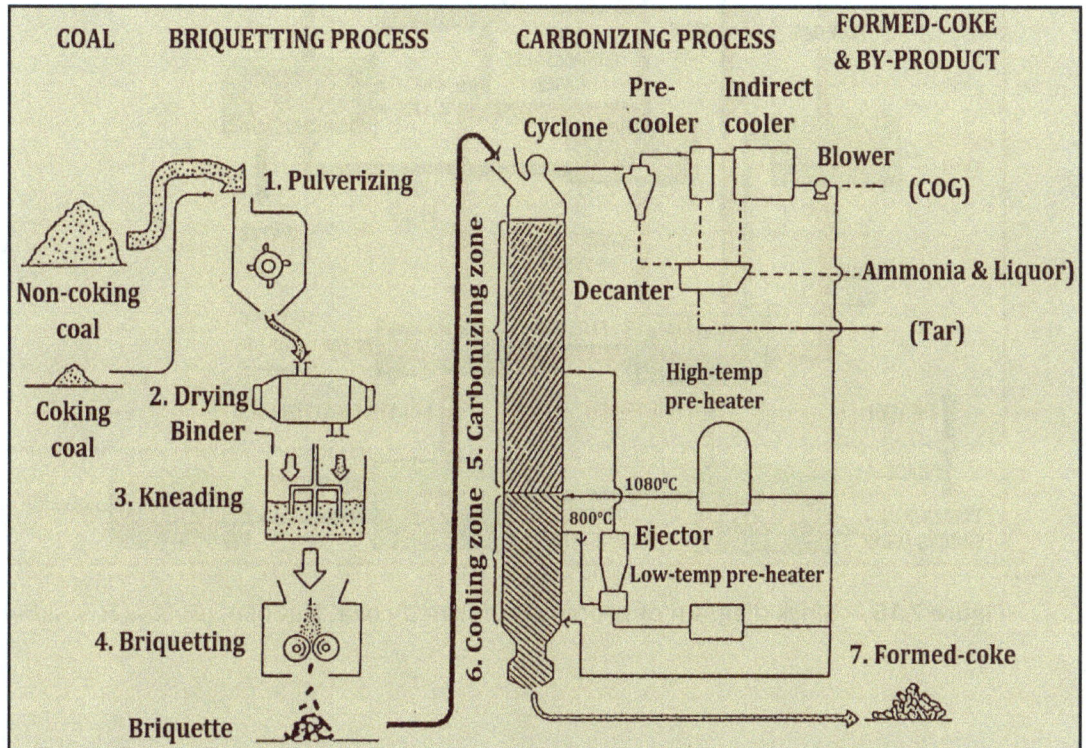

Figure 7.15   Flow diagram of the JISF continuous formed-coke plant.

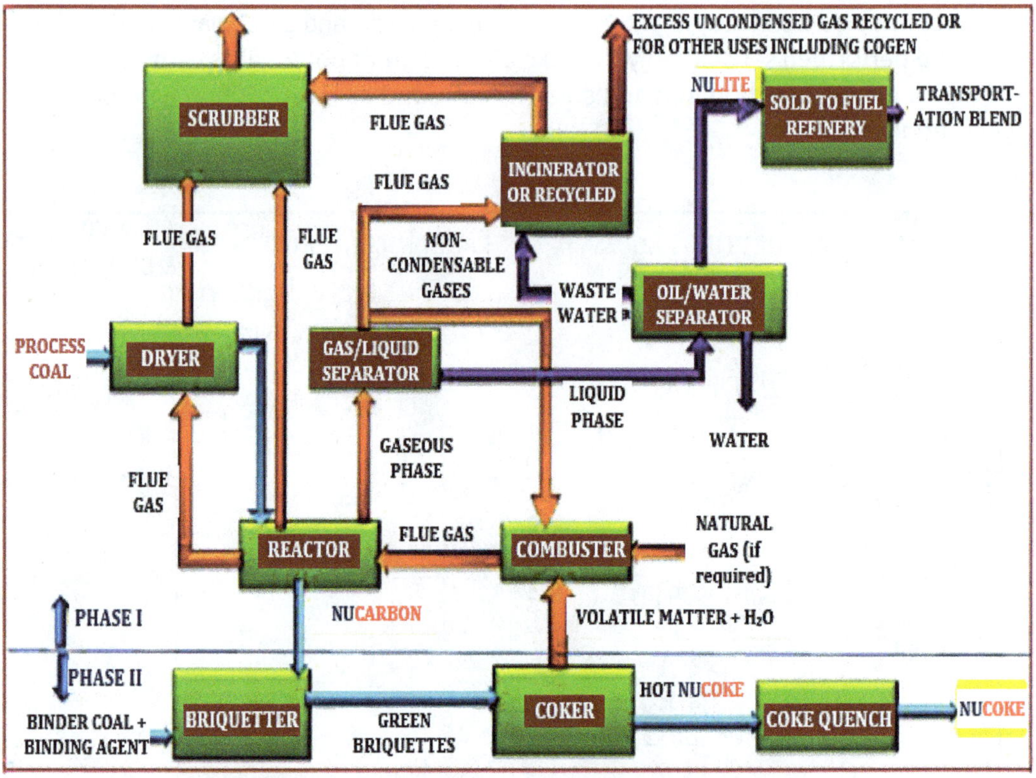

Figure 7.16   Block diagram of process for formed-coke. NuCoke. *(Cordova et al., 2014).*

Table 7.6. Characteristics of NuCoke and conventional coke.
(Cordova et al., 2014).

PROPERTIES	NUCOKE	CONVENTIONAL COKE
Moisture (%)	2 max	5-7
Volatile matter (%)	0.5-1.0	1.0 max
Ash (%)	7	8
Fixed carbon (%)	92	91
Sulphur (%)	0.6	0.7
Bulk dens ft.)ity (lb./cu.	38	29
Stability (%)	61-66	58
Hardness (%)	69	67
CSR, NSC (%)	65-74	55 min
CRI, NSC (%)	24-31	32 max

7.5.2   Production of formed-coke from Nigerian coals.

The Ajaokuta Steel plant adopted the iron blast furnace process and coke ovens with an annual capacity of about a million tonnes of coke are in place. However, none of the Nigerian coal deposits has good coking properties and about 90% of the annual requirement of 1.2 million tonnes (first-phase) of coking coal will have to be imported. The location of the plant does not favour easy transportation of imported coking coal since there is no rail line linkage with any sea port. The coking coal problem has been one of the major factors which delayed the take off of the plant for over forty years. Preliminary investigations discussed in Section 7.4 had shown that strong briquettes could be produced from Enugu, Okaba and Obi/Lafia coals without the use of a binder but the tests were carried out in small-scale laboratory equipment. The compaction was carried out in a heated cylindrical mould because it was difficult to control char temperature between exit from the fluidized bed and the briquetting unit of the equipment. However, the quality of the product was high. The crushing strength of the briquettes was an average of about 12,000kN/m^2, nearly twice the strength of a typical oven coke. Porosity was only about 60%, reactivity about double, while residual volatile matter was about three times the corresponding values for a typical oven coke.

Coke reactivity is largely a function of the residual volatile matter and affects indirectly the combustion in the blast furnace. The relatively high reactivity of the briquettes is undesirable but can be controlled at the secondary carbonization stage by adjusting the process parameters to obtain the optimum residual volatile matter. Also, it is to some extent compensated for by the loss of reactivity due to its lower porosity and more regular shape. The results of the small-scale briquetting tests were sufficiently positive to justify further tests on pilot-plant scale, the primary focus being on the production of formed-coke of metallurgical quality. The equipment is shown in Figure 7.17 and has been described fully elsewhere (Afonja, 1977).

The fist series of tests was designed to establish the critical variables - fluidization conditions, residual volatile matter and plasticity. Tests carried out in the laboratory fluidized-bed indicated an optimum fluidization temperature range of 400 - 470^{0}C and a residence time of 15-20 minutes. However, preliminary tests on the pilot plant established a minimum of 500^{0}C to achieve stable conditions. Further tests were carried out on Enugu, Okaba and Obi/Lafia coals at this temperature and various residence times as shown in Figure 7.18. For all three coals, residence time was significant for the first 20 minutes after which the residual volatile matter remained fairly constant. This contrasts with the conclusion of Kirov and Peck (1970) from tests carried out between 10 and 1000 minutes at fluidization temperatures between 300 and 1000^{0}C. The authors identified fluidization temperature as the only important factor and established a relationship as shown in Equation 7.11.

$$\text{Log } R(V) = 10.34 - 3.41 \log T \qquad\qquad [7.11]$$

The result of tests carried out on laboratory equipment similar to Kirov and Peck's also identified fluidization temperature as the only significant variable affecting residual volatile matter. It is clear however that residence time in the pilot plant is significant between 10 and 20 minutes (Figure 7.18). The difference may be explained by the fact that it would take more time to establish stable conditions in the pilot plant.

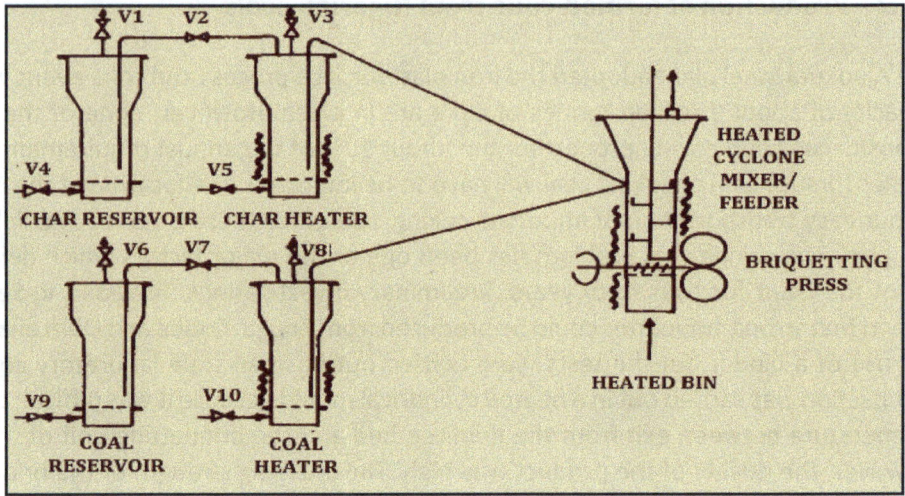

Figure 7.17 Schematic diagram of the formed-coke pilot plant facility.
(Afonja et al., 1977)

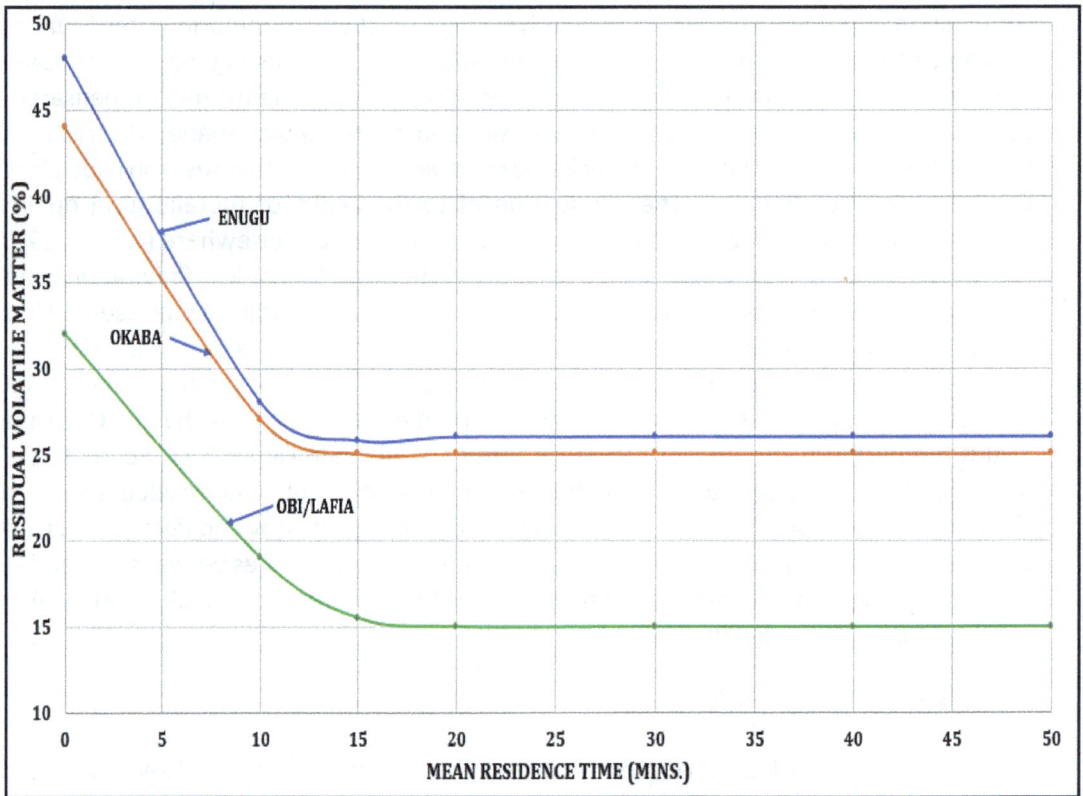

Figure 7.18 Effect of residence time on residual volatile matter of Nigerian coals at fluidization temperature of 500°C in Pilot Plant fluidized-bed. *(Afonja, 1979).*

More comprehensive tests were run on the pilot plant equipment to determine the optimum parameters for the production of formed-coke of metallurgical quality from either Enugu or Obi/lafia coal, or a blend of both coals. Each coal was ground to minus 2mm and stored in the coal reservoir unit of the pilot plant. Metered quantities were transferred into the fluidized-bed char heater, using pre-heated nitrogen as the fluidizing medium. The char was retained in the bed at a preset temperature in the optimum residual fluidity range for a specific period, and then evacuated into the cyclone-briquetter unit. Briquettes were formed without the use of a binder. Some samples of the char were taken prior to briquetting for volatile matter and residual plasticity analyses. The tests were repeated for various peak fluidization temperatures and residence times.

A second series of tests involved briquetting Enugu char-binder mixtures in a heated cylindrical mould on a Carver press. The char was prepared as described above, at 440°C and a residence time of 20 minutes using nitrogen as the fluidizing medium. The char was sieved to obtain the desired size consist and then re-heated in the bed. At a selected temperature, pulverized binder (coal tar pitch from Enugu coal) was injected into the middle of the coal bed and, after a short mixing time, the mixture was evacuated into the briquetting machine.

The third series of tests carried out on the pilot plant involved briquetting of char prepared from Enugu coal mixed with uncarbonized Obi/Lafia coal as binder. Controlled amounts of char and coal were transferred from the reservoir to the respective heaters and fluidized with nitrogen. The char was heated to about 750°C while the coal heater temperature was limited to 200 to 250°C to avoid agglomeration. At the appropriate temperatures the coal and char were blown into the the fluidized bed mixer maintained at a temperature between 350 and 450°C. The mixture was homogenized by stirring and fluidizing with nitrogen for about 60 seconds after which the material was fed to the briquetting press by a screw feeder. Ovoid shaped briquettes weighing about 20 grams were produced. The size range of the char was 100 to 800 microns. Preliminary test runs had shown that good fluidization, good material transfer, and minimum material loss through elutriation were achieved if the bulk of the char was in this size range. Also, the results of the laboratory tests had shown that the binder coal should be ground as fine as possible in order to promote intimate mixing and produce strong briquettes, hence the Obi/Lafia coal was ground to minus 200 microns. The main parameters of interest in these tests were char : binder ratio and size consist of the binder.

All briquettes were immersed in fine sand and carbonized for 2 hours in a 7kg Carbolite electric furnace at temperatures between 400 and 900°C in order to establish the optimum secondary carbonization temperatures. The ovoid-shaped briquettes from the pilot plant tests were processed in a Micum drum to determine the micum strength and abrasion resistance indices.

The optimum range of chloroform extract (0.8-1.2) was achieved at slightly higher temperatures compared with values established in the laboratory-scale equipment. Peak temperature of 500°C at a residence time of 15 minutes for Enugu and Okaba coals but a longer residence time of 20 minutes was required for Obi/Lafia coal. The results of the investigation showed that it is possible to produce good quality formed-coke from Enugu and Obi/Lafia coal chars prepared under conditions established for optimum residual plasticity, without the use of a binder. The optimum residual plasticity range as determined by the chloroform extract is achieved when Enugu coal is fluidized at temperatures between 420 and 440°C. The equivalent range for Obi/Lafia coal is 440 and 460°C. The Micum (M_{40}) index and crushing strength of formed-coke depend significantly on the peak fluidization temperature at the char production stage (Figures 7.19 & 7.20). The optimum temperatures fall in the established temperature range for optimum residual fluidity for both coals (Figure 20). There was a sharp deterioration in strength for holding times of more than about 60 seconds (Figure 7.21). This was due apparently to the fact that extended fluidization caused partial or total elimination of residual fluidity. Chars produced outside the identified optimum temperature range produced weak formed-coke (see Figure 7.19). At lower temperatures the residual volatile matter is unacceptably high and rapid and copious degasification during secondary carbonization will cause multiple fracture in the formed-coke, resulting in low strength. On the other hand, higher temperatures will not facilitate retention of the desired residual plasticity which is necessary for binderless briquetting of the char. Both the carbonization temperature and heating rate at the secondary carbonization stage are important (Figures 7.22 & 23). A high carbonization temperature is necessary to achieve high crushing strength. The highest crushing strength is achieved at the lowest heating rate. The influence of heating rate is more marked at the higher carbonization temperatures. The residual volatile matter of good, strong formed-coke should be no more than 2-5% and a secondary carbonization temperature of at least 900°C for both Enugu and Obi/Lafia coals (Figure 7.24).

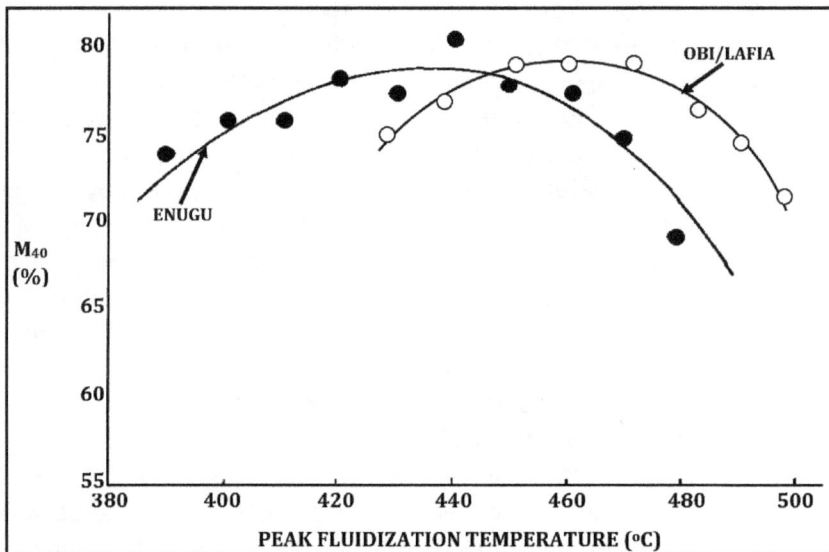

Figure 7.19   The effect of peak fluidization temperature of Enugu and Obi/Lafia chars on the strength of formed-coke. Secondary carbonization temperature = 900°C. *(Afonja, 1983).*

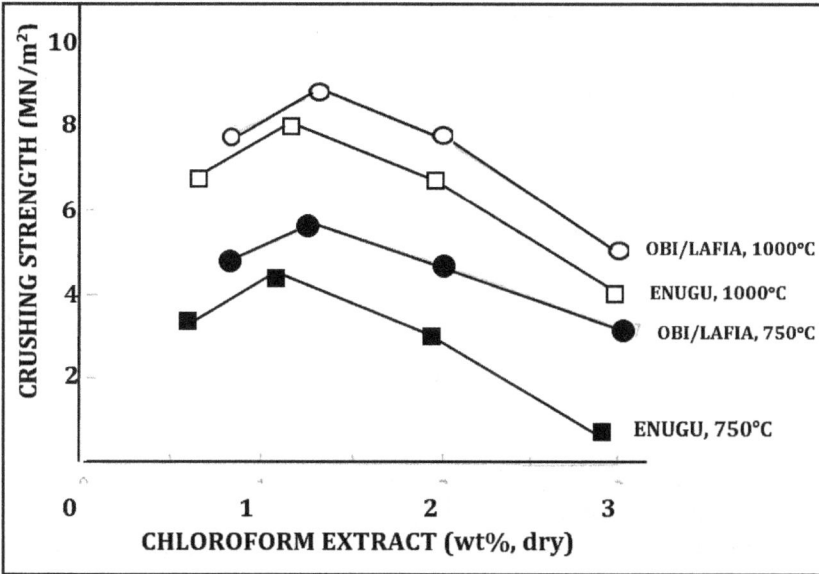

Figure 7.20 Effect of residual raw briquette plasticity as indicated by the chloroform extract yield and secondary carbonization temperature on the crushing strength of briquettes made from Enugu and Obi/Lafia coals without a binder. *(Afonja, 1986).*

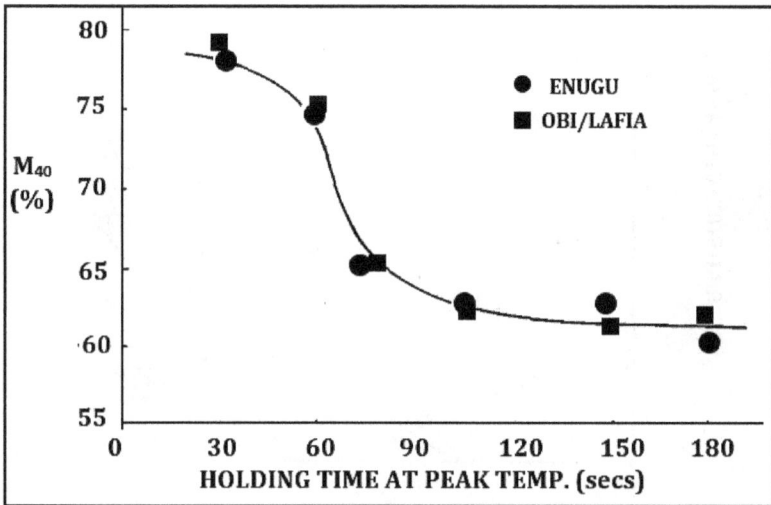

Figure 7.21 The dependence of the strength of formed-coke made by binderless briquetting of Enugu and Obi/Lafia char on the holding time at fluidizing temperature of 440°C. Secondary carbonization temperature = 900°C. *(Afonja, 1983).*

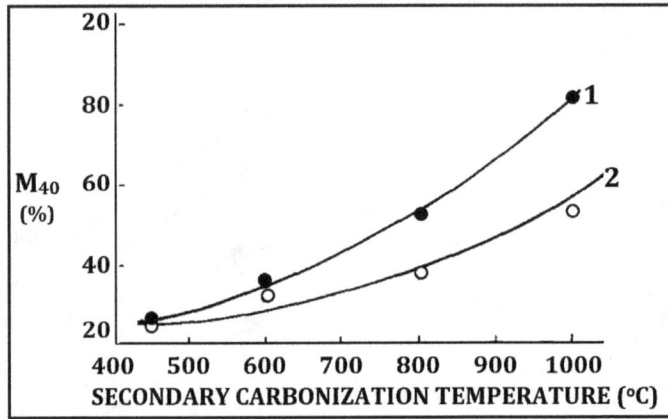

Figure 7.22 The effect of secondary carbonization temperature and heating rate on the strength of binderless formed-coke made from 440°C Enugu char. (1) heating rate, 3°C/minute; (2) heating rate, 10°C/minute *(Afonja, 1983).*

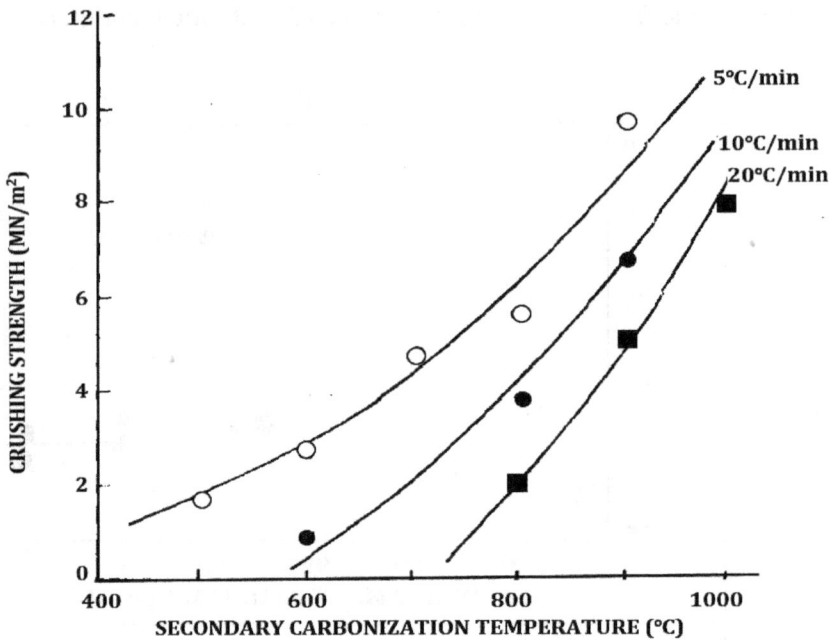

Figure 7.23 Effect of secondary carbonization temperature and heating rate on the crushing strength of briquettes made from Obi/Lafia coal by binderless briquetting. Fluidization temperature = 450°C, residence time = 20 minutes *(Afonja, 1986).*

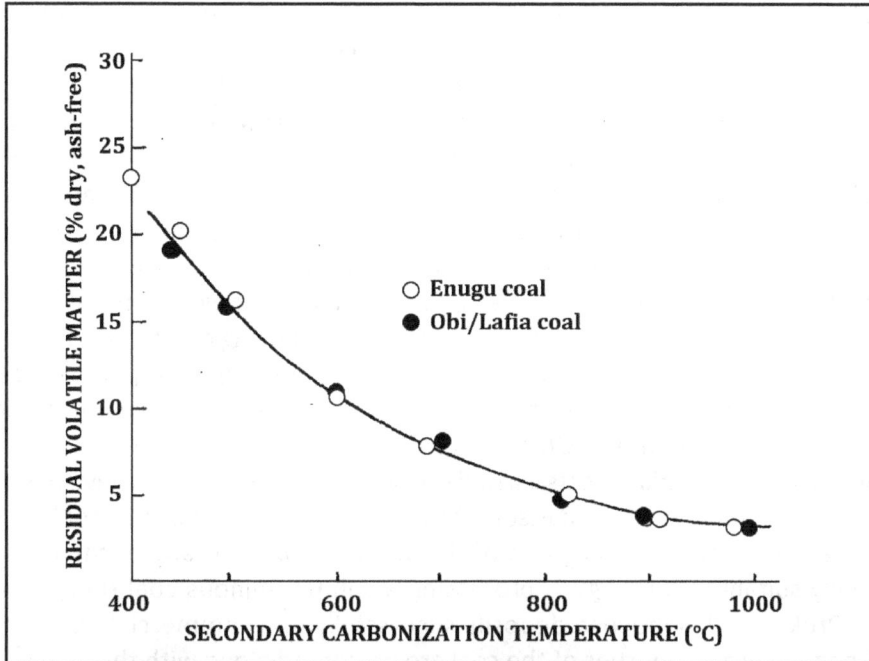

Figure 7.24 The dependence of residual volatile matter of binderless briquette made from Obi/Lafia and Enugu coals on the secondary carbonization temperature. Fluidization temperature = 450°C, residence time = 20 minutes *(Afonja, 1979; 1986).*

A sample of the formed-coke produced by briquetting Enugu Char produced under established optimum conditions without the use of a binder is shown in Figure 7.25. Good quality formed-coke can be obtained by briquetting char prepared from Enugu coal mixed with a binder (pitch or coking coal). The pitch used was obtained by processing the tar condensed from the flue gas produced at the fluidization stage. Obi/Lafia coal was also used as a binder but it had to be ground very fine. The particle size distribution of the char-coking coal binder is also critical.

A blend which consists of mostly fine particles yields briquettes of low strength. The particle size of the char has a significant influence on the ultimate strength properties of the formed coal (Figure 7.26). The highest crushing strength is achieved from a blend consisting of about 50% minus 600 microns. Also, briquettes made from coarse material within the optimum size range are considerably stronger than those made from fine material. The results of the laboratory and pilot plant tests showed that briquette strength improved with increasing Obi/Lafia binder coal content (Table 7.7). However, an increase in the binder coal particle size consist had a deleterious effect on crushing strength irrespective of the proportion. The tests also confirmed the positive effect of binder coal proportion on the strength (M_{40}) and abrasion index (M_{10}) of the carbonized briquettes. At least 10% pitch or 40% coal binder is required to obtain blast furnace quality formed-coke. However, in the case of coal binder it should be ground as finely as possible. The char used in the tests was prepared from Enugu coal under the optimum conditions established for formed-coke as described earlier.

In summary, comprehensive tests on laboratory equipment and pilot plant have shown that formed-coke of metallurgical quality can be produced by binderless briquetting of chars prepared from Enugu and Obi/Lafia coals provided the char is prepared in the fluidized bed at optimum fluidization temperature and residence times which ensure that the char retains optimum residual plasticity. The char should be heated to a temperature of about 440-460°C, the temperature range at which it becomes plastic. Other variables which affect the strength of the formed-coke include the secondary carbonization temperature and heating rate, and the residual volatile matter of the char. Good quality formed-coke can also be made from char prepared from Enugu coal mixed with soft pitch (ring and ball softening point of 80°C) or uncarbonized Obi/Lafia coal. When pitch is used, it should be mixed with hot char at a temperature well above the melting point of the pitch. Coking coal binder should be pulverized and the blend heated to its plastic range before briquetting. The particle size range of the blend appears critical in producing strong formed-coke. The optimum particle size range established for the Enugu char-Obi/Lafia coal blend was about 50% below minus 600 microns.

Although all formcoking tests carried out so far on Nigerian coals have been on laboratory of pilot plant equipment, comparison of available data with those of coals being processed in commercial plants in other parts of the world shows that any of the proven deposits is potentially suitable. NuEnergy is processing a sub-bituminous coal (PRB) similar to Enugu, Okaba, Orukpa and Ogboyoga deposits successfully in a commercial plant shown in Figure 7.27. The technical properties of the coal are presented along with those of Nigerian coals in Table 7.8. The NuEnergy process produces formed-coke from charred non-coking coal mixed with coking coal and pitch binders in the ratio 60:30:10. Obi/Lafia coal would have been a good substitute for the coking coal but the sulphur content is unacceptable. However, the results of pilot plant tests presented earlier indicate that coking coal could be eliminated by increasing the pitch content to around 15%. The pitch is a by-product of the charing process. It should be noted however that the moribund Ajaokuta Steel Plant would have been the major user of formed-coke. However, the product has good export potential.

Figure 7.25. Sample of formed-coke prepared from Enugu coal chared in a fluidized-bed at 440°C for 20 minutes, without the use of a binder. Secondary carbonization temperature was 900°C and heating rate of 3°C/minute *(Afonja, 1983)*.

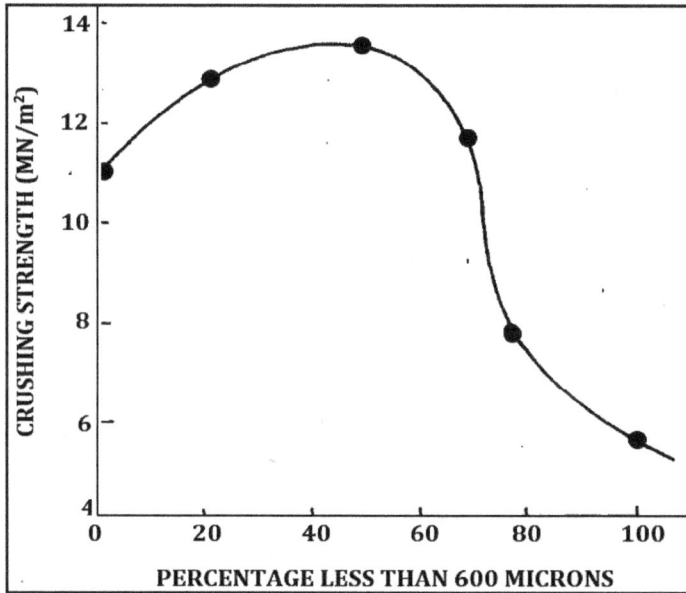

Figure 7.26 Effect of particle size consist on the crushing strength of Enugu char-Obi/Lafia coal briquettes carbonized at 900°C. Moulding temperature = 380°C. *(Afonja, 1986).*

Table 7.7 Strength and abrasion indices of briquettes made in laboratory and pilot plant tests. *(Afonja, 1986).*

SIZE CONSIST (mm)		STRENGTH/ABRASION INDICES OF CARBONIZED BRIQUETTES				
CHAR	COAL	CHAR/PITCH RATIO, LAB. TESTS		CHAR/COAL RATIO, PILOT PLANT TESTS		
		90/10	95/5	80/20	70/30	60/40
< 0.2=5.6% 0.2-0.39=27.1% 0.4-0.59=32.4% 0.60-0.80=34.9%	0-0.062=47.6% 0.063-0.10=52.4%	$M_{40}=85$ $M_{10}=6$	$M_{40}=68$ $M_{10}=9$	$M_{40}=75$ $M_{10}=6$	$M_{40}=75$ $M_{10}=6$	$M_{40}=75$ $M_{10}=6$
	0.063-0.099=42.3% 0.10-0.199=37.9% 0.20-0.50=19.8%			$M_{40}=67$ $M_{10}=12$	$M_{40}=74$ $M_{10}=11$	$M_{40}=77$ $M_{10}=10$

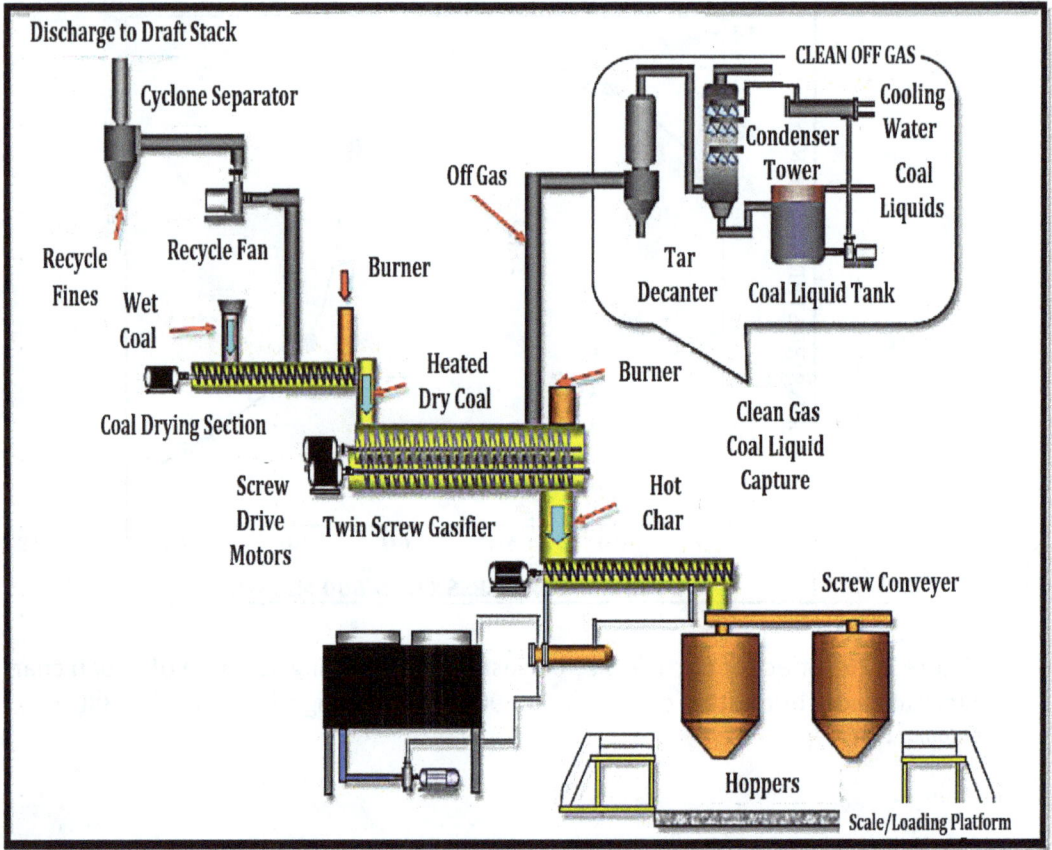

Figure 7.27 The NuEnergy continuous coal charing process. *(Cordova et al., 2014).*

Table 7.8 Technical properties of Nigerian and NuEnergy coals.
(Cordova et al., 2014).

Chemical Parameter	Coal Source					
	Enugu	Ezimo	Orukpa	Okaba	Ogboyoga	NuENERGY PRB
Moisture (%)	2.5-12.2	11.1-13.8	10.8-17.2	7.9-12.7	12.6-15.9	27.1
Ash (%)	4.1-25.5	5.9-13.1	7.1-12.3	6.8-11.1	8.1-9.5	9.5
Vol. Matter (%)	31.6-42.1	38.7-42.1	34.9-38.1	39.8-43.2	37.9-43.1	45.8
Fixed Carbon (%)	32.9-47.1	36.7-42.3	38.7-41.7	37.9-42.5	37.6-39.4	44.7
Carbon (%)	62.1-69.6	59.9-64.1	57.5-59.7	56.4-61.2	51.8-58.9	
Hydrogen (%)	4.1-5.3	1.2-4.9	3.4-3.9	3.5-3.8	3.6-4.9	
Nitrogen (%)	1.1-2.3	1.4-1.7	1.1-1.8	1.4-1.8	1.3-1.6	
Sulphur (%)	0.4-1.8	0.5-0.9	0.2-0.5	0.5-0.8	0.4-0.8	0.35
Oxygen (%)	8.2-14.1	11.0-13.1	10.9-14.4	14.9-17.9	14.8-21.1	
Calorific Value (MJ.kg)	22.1-28.7	24.6-25.8	25.1-26.1	22-7-24.6	22.8-24.1	26.6

REFERENCES

Aderibigbe and Szekely, (1982). "Studies in coke reactivity: Part 2, Mathematical model of reaction with allowance fore pore diffusion and experimental verification." *Ironmaking and Steelmaking*, Vol. 9, No. 1, pp. 32-40.

Afonja, A. A. (1972). "An assessment of the coking quality of Nigerian coal seams." ENE/B/W/3. Report No. IFE/CHE/CP/1. For the Nigerian Steel Development Authority. (N.S.D.A.).

Afonja, A. A. (1973). Preliminary studies on the briquetting of Nigerian coals. *Institute for Briquetting and Agglomeration*, pp. 265-271.

Afonja. A. A. , Lau, I. T. and Leeder, W. R. (1977). "A formed-coke Pilot Plant Facility." *International Institute for Briquetting and Agglomeration*,14, p. 239.

Afonja, A. A. (1979). "Further Studies on the Obi/Lafia Coal Deposit (Phase II)." Research Report No. IFE/CHE/CP/11, Coal Research Laboratory, Ile-Ife.

Afonja, A. A. (1981a). "Optimization of the utilization of a sub-bituminous coal in conventional and form-coking." *Fuel Processing Technology*, 7, pp. 293-310.

Afonja. A. A. (1981b). Fluidized-bed Carbonization of Nigerian Coals. *Proc. Nig. Soc. of Chemical Engineers*, p. 50.

Afonja, A. A. (1983). "A study of the potential of Nigerian coals for metallurgical coking." *Proc. 1st National conference on steel*, Ovwian-Aladja, April 18-20, pp. 51-95.

Afonja, A. A. (1986). "A Study of the Potential of Low Grade Coals for the Production of Blast Furnace Coke." *Proc. International Iron & Steel Congress*, Washington D. C. pp. 239-246.

Afonja, A. and Olofinjana, O. (1988). *Formed-coke production from local materials for blast furnace utilization.*, Report No Ife/CHE/CP/16. Submitted to Nigerian Steel Development Authority. (See also Olofinjana, O. M.Sc. Thesis, OAU, Ile-Ife, 1987).

Afonja, A. A. (1996b). " Production of metallurgical coke from non-coking coals." *In Nigerian Coal: A Resource for Energy and Investments*. Eds. H. Okolo and M. Mkpadi, pp. 89-109. Raw Materials Research and Development Council, Nigeria.

Afonja A. A. (2017). *Basic Coal Science and Technology*. SineliBooks.

Ahland, E., Bock, B. Jagnow, H. J. and Lehman, J. (1971). "Hot briquetting processes for the production of fuel for the metallurgical industry." *Proceedings of the 12th Annual Conference of the Institute of Briquetting and Agglomeration*, Vancouver, Canada.

Arthur D. Little Inc. (1965). *Revitilizing the Enugu Coal Industry Through Large-Scale Carbonization*. Duffryn Technical Services Ltd. Chemical Utilization Project Report to the Under Secretary of State for the colonies, London.

Bell, D., Towler, B., and M. Fan (2011). *Coal Gasification and its applications*. Elsevier, Burlington. MA.

Brown, K. (2012). "In situ gasification: An emerging technology." www.asmr.us/Publications/Conference Proceedings/2012/

Dartnel, J., (1978). "Formed-coke – still waiting in the wings.: *Iron and Steel International*, Vol. 51, p. 155.

de Cordova, M., Walia, D., Madias, J., Cornejo, A. (2014). "NU ENERGY Cokemaking Technology for Current Challenges." *AISTech 2014 Proceedings*, pp. 305-314.

ETSAP (2010). "Syngas production from coal." Energy Technology Systems Analysis Programme, www.etsap.org

Gregory, D. H. (1966). "formed-coke from high volatile coals." Proc., *Congress International   de Charleroi*: A6.

Higman, C., and S. Tam (2013). "Advances in Coal Gasification , Hydrogenation, and Gas Treating for the Production of Chemical Fuels." American Chemical Society.

Higman, C. (2014). "State of the gasification industry: worldwide gasification database 2014   update." Gasification Technologies Conference, Washington D.C., 29th December, 2014.

Jensen, J., Poulsen, J. and N. Andersen (2011). "Substitute natural gas." Nitrogen+Syngas 310 (March-April, 2011).

Kohno, 1988) Kohno, T. (1988). Substitution and reduction of coking coal in cokemaking. Coal and Coke Technical Exchange Session, International Iron and Steel Institute Committee on Technology, Brussels, April 1988., pp 85-101.

Nekrasov, Z. I., Ketov, K. I., Gladkov, N. A., Zhembus, M. D., and Goncharov, V. F. (1974). Some results from test melts with formed-coke. Steel in the USSR, Vol. 29, pp 556-568

Philips, J. (2015). "Different types of gasifiers and their integration with gas turbines." www.netl.doe.gov

Sasol-Lurgi Technology (2005). (Sasol-Lurgi Coal Gasification Technology and Low-Rank Coal." The Proceedings, Gasification Technologies Conference, 10-12 October, 2005, San Francisco.

Wang, T., and G. Stiegel (2017). *Integrated Gasification Combined Cycle (IGCC) Technologies.* Elsevier.

8 Conversion potentials of Nigerian coals: premium chemicals and materials

8.1 INTRODUCTION

Coal tar was probably the first chemical obtained from coal in the 1700s and was used extensively in shipbuilding in England. By the mid 1850s, lamp oil and fuel gas were being produced from coal in different parts of the world. The development of the iron and steel industry in the 19th century made coal tar widely available as a by-product of the coke oven. A process for tar distillation was developed in Germany in 1917 and produced aromatics for use around the world for the production of pharmaceuticals, dyes, explosives, and photographic chemicals. Germany's embargo on aromatic and coal tar products during World War 1 prompted the development of coal-based chemical industries in other countries, in particular, England and the USA. Germany had also developed a process of producing liquid fuels from coal in the 1920s and several commercial plants were producing gasoline, diesel, aviation fuel, etc. from coal by the 1930s.

South Africa acquired the coal-to-liquid fuel technology, also known as "Fischer-Tropsch" (F-T) technology from Germany and further developed the process to convert the country's vast coal resources to liquid fuels. There were parallel efforts in several other countries to develop different coal-to-liquid fuel technologies. However, coal conversion to chemicals declined from the 1950s due to the increasingly wide availability of oil and gas, but development continued in South Africa. The F-T technology has re-emerged in recent decades due mainly to the instability of the crude oil market and the desire of many developed countries to reduce dependence on foreign oil sourced mainly from an unstable region.

8.2 COAL-TO-LIQUID FUELS

Coal can be converted to liquid fuels that can be utilized as an alternative to oil. There are two different methods: *Direct coal liquefaction* and *indirect coal liquefaction* (Figure 8.1). The indirect process has been in commercial operation since the 1930s in Germany and produced petrol and diesel oil for the German World War II effort. However production in Germany ceased after the war and, with the increasing availability of relatively cheap oil, most other countries lost interest in the technology. On the contrary, South Africa acquired the technology under licence from Germany to convert the country's vast coal resources to fuel oil.

This explains why most of the plants in commercial operation presently are located in South Africa. Also, the first stage of the process which gasifies coal has become a standard unit in many coal-fired power plants all over the world. The direct process was developed in Germany, the United States, and United Kingdom as an alternative to the F-T technology in the late 1930s to early 1960s but interest declined with the wide availability of petroleum fuels in the 1950s. The instability in the global oil market in the early 1970s prompted a resuscitation of research and development activities in the USA, Germany, Japan and other countries The focus has been the reduction of the cost which was considered prohibitive. However, to date, only one commercial plant is in operation anywhere in the world presently and it is located in China.

8.2.1   Direct coal liquefaction (DCL)

In the DCL process, coal is dissolved in a solvent and reacted with hydrogen (hydrogenation) at high temperature and pressure in the presence of a catalyst (Figure 8.2). The raw liquid product is further refined into product liquid fuels. The first generation DCL processes featured a single reaction stage that liquefies coal, possibly with an integrated on-line hydrotreating reactor to upgrade the distillates. The more recent DCL processes from the 1970s involve two stages. In the first reactor coal is dissolved in a solvent. The high molecular weight product is upgraded by catalytic hydrogenation into lower-boiling liquids with reduced heteroatom content. The DCL process is referred to as direct because the coal is transformed into liquid without being first gasified. In principle, this process is highly efficient because it is simpler and causes less atmospheric pollution compared with the indirect process. Furthermore, the hydrogen required is produced by gasifying coal to produce syngas which is further treated to produce hydrogen. However, the technology is relatively more expensive.

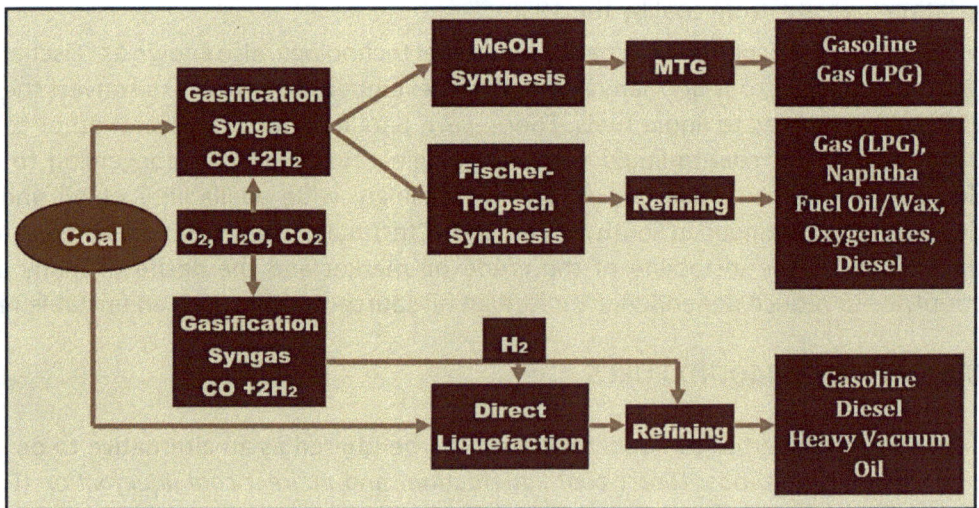

Figure 8.1 Options for coal liquefaction. *(ExxonMobil, 2014).*

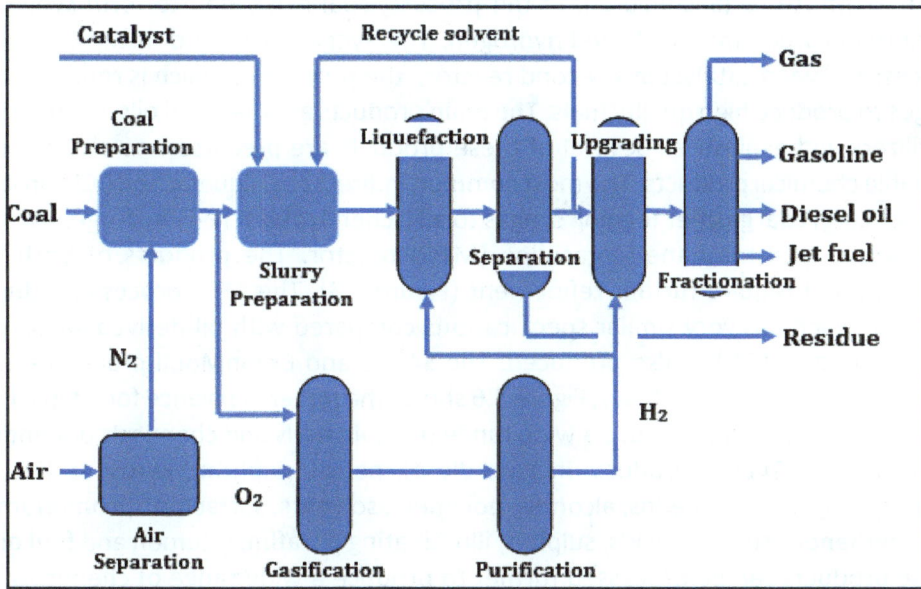

Figure 8.2 Shenhua direct coal liquefaction process. *(Xiuzhang, 2010).*

The DCL process produces mainly aromatics and a wide range of other chemical compounds of varying molecular weights, hence substantial upgrading is required to yield transportation fuels of acceptable quality. A comprehensive review of the current status of the DCL process has been published by Vasireddy et al., 2011; Shui et al., 2011; Mochida et. Al., 2014). One of the major assets of the direct coal liquefaction process is the ability to process any grade of coal to high quality liquid fuels. The process is particularly effective for low-grade coals. Lignite coal (2000-4000 kcal/kg), and sub-bituminous coal (4000-6000 kcal/kg) can be processed to high energy density liquid fuel (10000 kcal/kg) which is distilled to obtain gasoline, diesel oil and kerosene. Processing higher energy density bituminous coals (6000 kcal/kg and above) is less cost effective.

A US process was licensed to Shenhua Corporation in China in 2002 and a demonstration plant began operation in Erdos, in the huge coal fields of Inner Mongolia in 2008 (Xinjua News, 2013; DOE, 2015). The Erdos plant which had an initial capacity of 20,000 barrels per day has been expanded and, since 2011, has been producing over 1 million tonnes of liquid products a year, including diesel oil, liquefied petroleum gas (LPG), and naphtha (petroleum ether). A New CTL plant owned by a Shenhua subsidiary went into production in 2016 in Northwest China's Ningxia Hui region. The plant which is the world's biggest has the capacity to process 20 million tonnes of coal a year, producing 4 million tonnes of products including diesel, naphtha petroleum, an liquefied gas. Other products include sulphur, alcohols and ammonium sulphate (China Daily, 2016)

8.2.2   Indirect coal liquefaction (ICL)

The indirect coal liquefaction process was developed by Lurgi originally in Germany in the 1920s and 1930s but most of the subsequent development has been in cooperation with

Sasol South Africa (Sasol-Lurgi). In the process, coal is gasified to form syngas which is a mixture of carbon monoxide and hydrogen. The syngas produced in the first stage is then processed over a catalyst in a second reactor, the product of which is refined in subsequent stages to produce high quality fuels. The main products are fuels and oils, chemicals, polymers, fertilizers and explosives. Several of these products are precursors to a very wide range of valuable chemical products. The most common indirect coal liquefaction (ICL) process features a Sasol-Lurgi coal gasifier feeding syngas to a Fischer-Tropsch (F-T) hydrogenation reactor, or the ExxonMobil methane-to-gasoline (MTG) reactor. The products of both options are different and require further refinement (Figure 8.3). The MTG process produces gasoline (petrol) which has very similar specifications compared with oil-derived gasoline. Liquefied petroleum gas (LPG) is also produced. The SASOL and ExxonMobil processes are shown in Figures 8.4 and 8.5 respectively. Figure 8.6 shows the material balance for a typical ICL process.

The F-T route can produce a wide range of liquid fuels and chemicals depending on plant configuration. Typical products include diesel, petrol, naphtha, kerosene (jet fuel), liquid petroleum gas (LPG), olefins, alcohols, polymers, solvents, surfactants, comonomers, ammonia, methanol, crude tar acids, sulphur, illuminating paraffin, bitumen and fuel oil. Several of these products can be processed further to produce a wide range of chemicals. Up to 2009 there was only one commercial coal liquefaction plant in the world, located in South Africa. Four demonstration plants have been established in China in the last few years and several projects are under development in Australia, India. Mongolia, Botswana, Russia and USA. The first MTG Coal-to-Liquid 2,500 barrels/day gasoline production demonstration plant began operating in 2009 in Shanxi province, China as part of a demonstration complex which includes coal gasification, gas clean-up and methanol synthesis (El Malki and Hindman, 2014). Several new and very large coal-to-liquid plants have come on stream in China in the last five years and some are under construction in other countries.

The direct coal liquefaction process (DCL) is a one-step, highly efficient process. It produces higher octane gasoline with high energy density (MJ/litre) than the indirect conversion (ICL) process. However, the products have high aromatic content and low-cetane diesel is produced. Furthermore, operating costs are higher compared with the ICL process and operating experience on commercial scale is relatively limited. Another problem with the DCL process is its inability to process all types of coal. By comparison, the ICL produces ultra-clean fuels from any coal at lower operating costs.

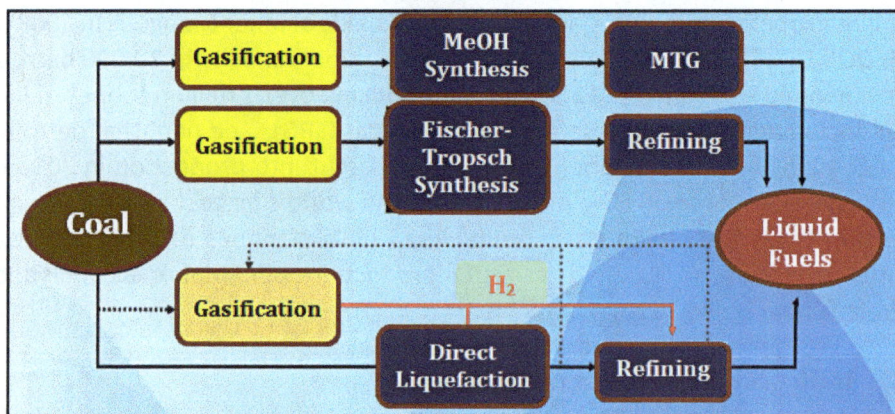

Figure 8.3 Alternative routes for processing syngas produced by the Fischer-Tropsch gasifier to liquid fuels.

The process is versatile and one gasifier can be configured with a liquefaction line and a power generating plant. Carbon capture is relatively easy and operating experience is extensive, gained from commercial plant operation dating back over six decades. Another major advantage of the ICL technology is its ability to process virtually any grade of coal. However, the ICL technology is a two-step process and therefore relatively inefficient. Furthermore, products have lower energy density. A recent development in coal liquefaction technology involved a hybrid DCL/ICL concept which seeks to combine the positive aspects of both technologies to produce high quality fuel at lower costs (DOE, 2014). A possible flow diagram is shown in Figure 8.7.

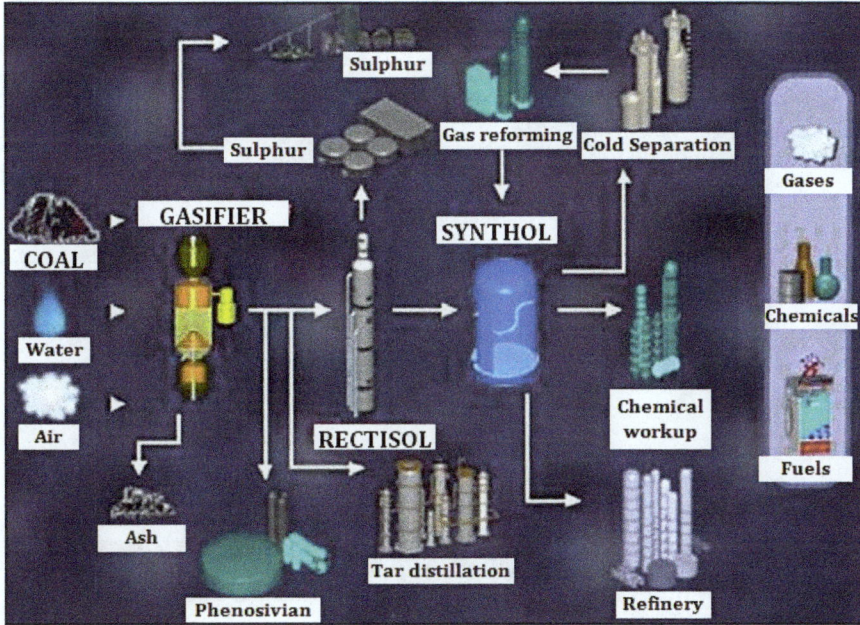

Figure 8.4 SASOL Indirect Coal Liquefaction processes.

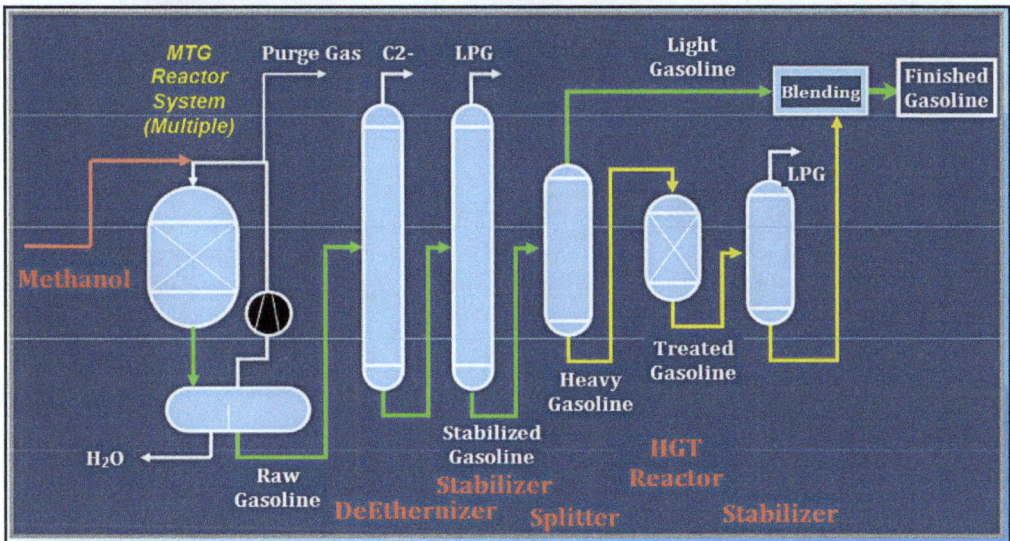

Figure 8.5 ExxonMobil MTG Coal to Liquid process.

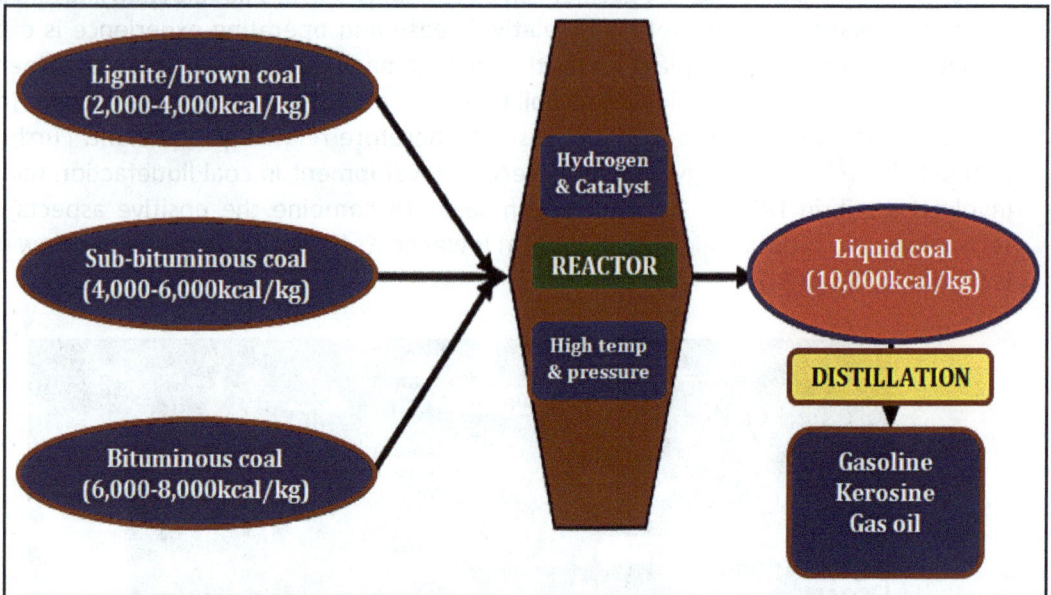

Figure 8.6 Materials balance of a typical Indirect Coal Liquefaction (ICL) process.

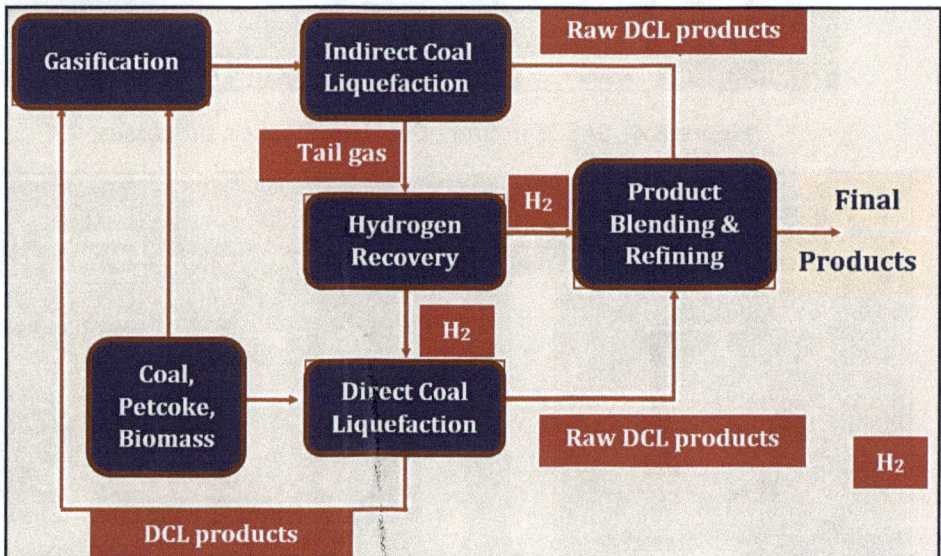

Figure 8.7 Schematic flow diagram of a hybrid DCL/ICL coal liquefaction process. *(Tennant, 2014).*

8.3   COAL CONVERSION TO CHEMICALS AND PREMIUM CARBON MATERIALS

Coal is a potential source of over a thousand valuable chemicals including ethanol, methanol, ammonia, calcium carbide, acetylene and hydrogen (Figure 8.8). This topic has been treated in some depth in another book (Afonja, 2017). Many of these products are feedstocks for the production of many other chemicals and materials including fertilizers, polymers, industrial dyes, and pharmaceutical products. Synthetic graphite from coal is also the precursor to high-technology materials including graphene, carbon nanotube, carbon fibre composites and high temperature lubricants. Industrial diamond produced from coal is used extensively in machining and jewelry.

Important polymeric and carbon materials derived from coal include a wide variety of engineering plastics and aromatic polymer materials, polymer membranes, liquid crystalline polymers, thermoplastics, high temperature heat-resistant polymers, graphitic carbon materials, carbon fibres, composites, molecular sieves, etc. Many of the starting precursors to these materials are not readily available from petroleum. For example, coal tar is still a major source of aromatic and heteroatom-containing chemicals which are converted to monomers for polymer synthesis. Furthermore, coal tar is still the source of nearly all the global demand for 2- to 4-ring polyaromatic and heterocyclic chemicals and a high percentage of BTX chemicals.

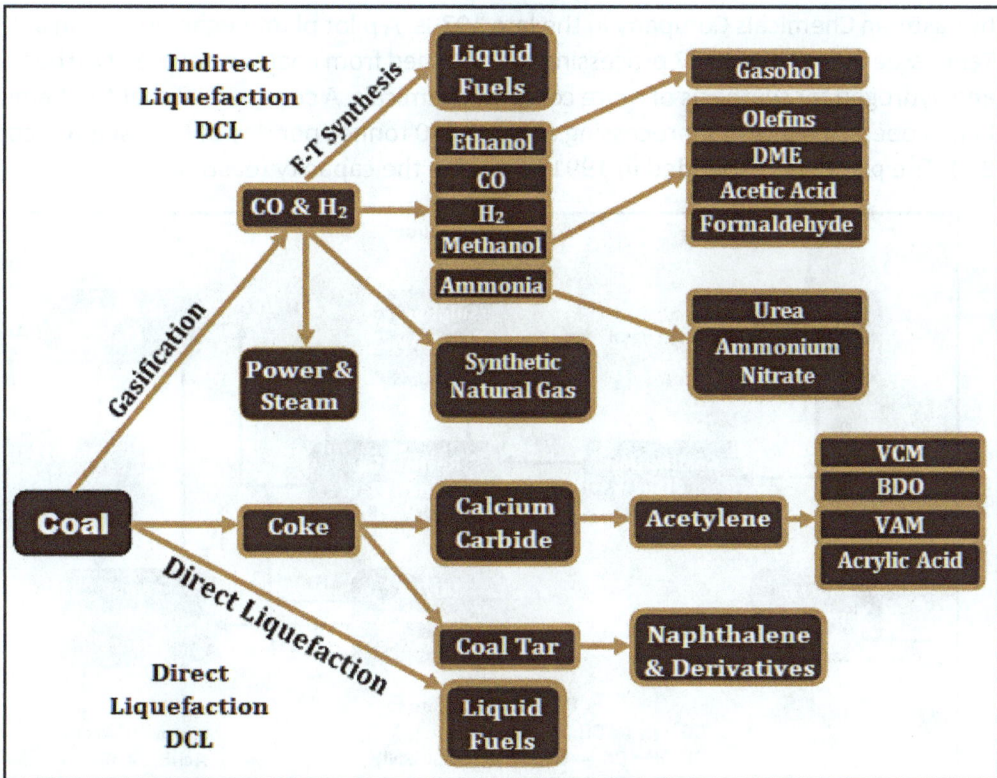

Figure 8.8   Coal To Chemicals conversion routes. *(Chemsystems.com).*

8.3.1 Coal conversion to chemicals

Coal has been a valuable source of chemicals for centuries. Coal tar and lamp oil distilled from coal were being used in Europe in the 1700s. In the late 1800s fuel gas and metallurgical coke were being produced from coal. The by-product gas of carbonization was being processed to obtain tar which was refined to obtain aromatics for pharmaceuticals, dyes, explosives and photographic chemicals. In the first half of the 20th century, coal was converted to gases or liquids, which typically were reacted with carbon monoxide in the presence of catalysts to form methanol and other similar chemicals. The widespread conversion of coal to acetylene via calcium carbide marked the beginning of a new phase in the development of the coal-to-chemicals industry. The seven basic organic chemical building blocks: ethylene, propylene, butadiene, benzene, toluene, the xylenes and methanol were all being produced from coal and this made possible the rapid development of the polymer industry in post World War II. It became possible to produce many chemicals on a large scale, including vinyl chloride monomer (VCM), acrylic acid (AA), acryonitrile (CAN) and 1,4-butanediol (BDO). Other products included olefins and acetylenes which were precursors to many other chemical products.

Production of coal chemicals declined in the 1950s due to the availability of cheaper petroleum and natural gas which now supplied around 80% of organic chemicals while only 20% was sourced from coal. Despite this development, coal-sourced chemicals have remained important, particularly in the polymer and carbon-based materials industries. The turbulence in the petroleum market in the early 1970s rekindled interest in coal-to-chemicals, pioneered by Eastman Chemicals Company in the late 1970s. A pilot plant began operation at Kingsport, Tennessee in the US in 1977 processing gas obtained from coal as a source of carbon monoxide and hydrogen for synthesis of more complex chemicals. A commercial coal-to-chemicals plant began operation in 1983, processing about 1,200 tonnes per day of high-sulphur coal (Figure 8.9). The plant was expanded in 1991 to double the capacity (eastman.com).

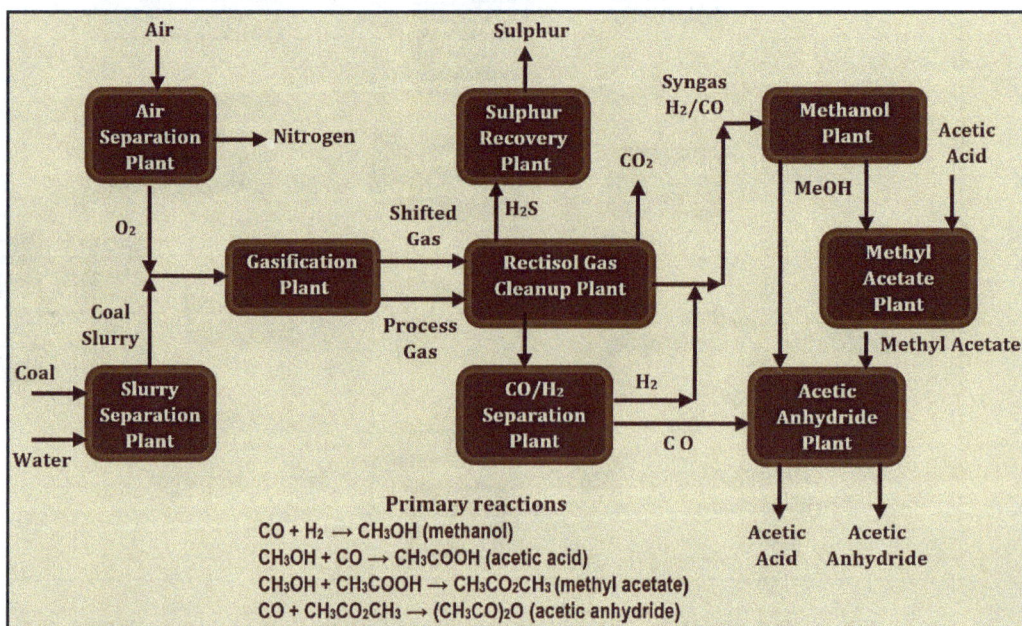

Primary reactions

$$CO + H_2 \rightarrow CH_3OH \text{ (methanol)}$$
$$CH_3OH + CO \rightarrow CH_3COOH \text{ (acetic acid)}$$
$$CH_3OH + CH_3COOH \rightarrow CH_3CO_2CH_3 \text{ (methyl acetate)}$$
$$CO + CH_3CO_2CH_3 \rightarrow (CH_3CO)_2O \text{ (acetic anhydride)}$$

Figure 8.9 Eastman coal-to-acetic acid process. *(eastman.com).*

In the Eastman coal-to-chemicals process, syngas from coal is purified and separated into hydrogen for the production of methanol, and carbon monoxide is processed into acetyl chemicals which are important building blocks in the synthesis of a wide range of consumer products, including plastics, textile fibres, and photographic film and chemicals. Hydrogen sulphide recovered from the purification process is converted into elemental sulphur and carbon dioxide is produced for the carbonated beverage industry. The main products are acetic anhydride, acetic acid and methyl acetate which are important precursors in the synthesis of a wide range of consumer products including pharmaceuticals, polymers, food containers, tool handles, auto coatings, auto interior components, performance fibres, specialty additives, personal care products, sweeteners, graphic arts inks, industrial films, displays, textiles, medical and opthalmic devices, bioplastics, thermoplastics, etc.

The Sasol-Lurgi coal-to-liquid technology has been in commercial operation since the 1950s and produces syngas which can be converted in an FT reactor to raw liquid fuel from which a wide range of liquid fuels and chemicals can be obtained. The focus of the Sasol process has been the production of automotive fuels but a range of basic chemicals is also produced for further processing to obtain valuable industrial chemicals. The MTG process produces methanol which is converted to gasoline. However, methanol is also a feed for the production of many chemicals.

The adoption of coal-to-chemicals technology has been fueled in the last decade or so by China's aggressive coal conversion policy (Minchener, 2011; Mochida et al., 2014; China Daily, 2016). In the last five years eight large coal gasification to chemical plants producing methanol, polypropylene, olefins and ammonia have been commissioned in China. As at 2017, China had 46 coal-to-chemicals plants and another 22 were projected to come on stream by 2020. The worldwide downturn in the coal industry in recent years has provided the impetus for the rapid development of coal conversion projects that produce more valuable and exportable products, in particular, auto fuels and industrial chemicals. China has the world's third largest proven coal reserves (after the US and Russia), holding 13% of the global total. However, China leads the world in both production and use of coal, consuming about 48% of the global demand. The bulk of this consumption is for power generation but the coal-to-chemicals sector has grown at a phenomenal rate in the last two decades.

About 80% of China's coal deposits are concentrated in the remote north-west region, up to 8000km from the industrial central and eastern regions where the coal is needed. A thrust of the Chinese coal-to-chemicals policy is to locate plants for conversion of stranded low-cost coal in the north-west to gas and high-value chemicals which are easier and cheaper to transport to the industrial regions. Coal- to-chemical plants are extremely carbon intensive and China is under heavy pressure to reduce CO_2 emissions (China currently leads the world). There are indications that this may slow down future plans for more coal conversion plants to some extent. However, coal-to-chemicals plants account for only about 3% of the country's total emissions and the impact may not be significant.

8.3.2   Coal conversion to premium carbon materials

Apart from fuels and chemicals, many high-value industrial materials are produced from coal. Carbon is an exceptionally versatile element, capable of forming a wide range of allotropes, from soft graphite to hard diamond, depending on the crystallographic arrangement of the atoms. Graphite and diamond, both allotropes of carbon, occur in nature but are believed to be of different origins. Graphite is the highest rank of coal. Like coal, it is formed from buried plant debris but there are significant differences in the geophysical conditions. Graphite is found in deep deposits in rocky areas and the unique structure compared with coals of other ranks is due to the exceptionally high pressure and temperature resulting from the high, rocky overburden. It is relatively rare, hence it is usually not considered as an energy source. Diamonds are believed to have been formed from carbon-containing organic minerals at exceptionally high temperatures and pressures, up to 200 kilometres beneath the Earth's surface over periods from 1 billion to 3 billion years ago. They are brought nearer to the Earth's surface by volcanic eruptions. The rarity and exceptional luster of diamond on polishing make it perhaps the most valuable gem in the jewelry industry.

Both diamond and graphite can be synthesized from coal tar. Synthetic diamond, also known as cultured diamond, is produced by several different methods but the most common are the high pressure and high temperature (HPHT) and Chemical vapour deposition (CVD) processes. Synthetic diamond is versatile, used extensively for cheap costume jewelry and industrial cutting blades and saws, machine cutting and drilling tool bits, abrasive and polishing grits and powders. Boron-doped synthetic diamond is also used in some electronic applications. Although the material is non-conductive, it can be made conductive or superconductive by doping with some elements, in particular, boron.

Graphite is a very versatile industrial material: it is used in high-temperature industrial furnace lining, foundry melting crucibles, electrodes for electric arc furnaces, fuel cell bi-polar plates, melting crucibles, manufacture of high-temperature lubricants, high temperature coatings, corrosion prevention coatings, electrolytic processes, conductive filters, carbon fibres, activated carbon, molecular sieves, rubber and plastic compounds, drilling applications, etc. Graphite can be synthesized by simulating the very high temperature and pressure conditions under which the material was formed naturally. Synthetic graphite can be manufactured by high-temperature treatment of amorphous carbon and precursors are many: any coal, coal tar pitch, petroleum coke, natural and synthetic organic materials (Asbury Carbons, 2006).

Most of the world's current graphite production processes use coal tar pitch or calcined petroleum coke as precursors. Both materials are composed of highly graphitizable forms of carbon. Synthetic graphite can be manufactured in varying degrees of technological specifications depending on application, and in any number of forms including solid articles of varied shape and size, granular and powder forms. Graphite is also a precursor to a wide range of premium carbon materials. These include calcium carbide, activated carbon, carbon fibre, industrial graphite, industrial diamond. High value nanotechnological materials such as nano carbon fibre, nanotubes, fullerenes and graphenes are also produced from synthetic, coal-derived graphite.

Graphite is made up of planar layers of covalent-bonded carbon atoms, each layer known as graphene. In effect, graphene is a one-atom thick planar sheet of graphite and materials made up of several layers of graphene are known as nanoplatelets. It is the strongest material to date and studies have shown that its breaking strength to weight ratio is around 100-300 times that of steel (Lee et al., 2008). Graphene conducts heat and electricity with great efficiency and is nearly transparent. The source of the strength of nanoplatelets is the exceptionally strong planar aromatic bonds. The material conducts heat and electricity better than any other known material, due to the incredibly fast rate at which electrons can move through the material. The 2-dimensional structure of graphene makes it possible for every atom to be available for chemical reaction from two sides. This makes graphene the most reactive form of carbon, around a hundred times more reactive than thicker sheets (Gabrielsen, 2013). The unique technological properties make graphene a building block for a wide variety of new materials, from structural carbon fibres, fulerenes, to electronics and biomedical materials. It is also one of the most studied new materials currently. The fact that graphene can be embedded in other matrices renders it exceptionally versatile because materials with combined strength and flexibility, exceptionally high thermal and electrical conductivity can be produced. However, like most other carbon materials, graphene has relatively low fracture toughness, only 10-20% of the range for metallic materials, depending on the impurity content.

8.4   FUTURE OF COAL-TO-LIQUIDS, CHEMICALS AND PREMIUM CARBON MATERIALS

There is little doubt that adoption of coal to liquid and chemicals technologies will grow rapidly in the near future, particularly in countries that have large coal resources but no petroleum resources. Various estimates indicate that coal liquefaction becomes economical when the local price of crude oil rises above $50-60 per barrel. However, economics is not the only determinant. Even countries that have large reserves of both oil and coal are actively developing or adopting coal conversion technologies as part of national energy independence policies. Also, coal conversion significantly mitigates the environmental degradation associated with by-products of coal-fired power generation because it is easier to capture the undesirable by-products and convert them to useful products. Furthermore, coal conversion plants can be located in coal mine complexes so that pollution is localized.

China is leading the world in coal gasification and chemicals from coal. Plans for 120 coal-to-chemicals plants have been announced and the main products will be liquid fuels, monoethylene glycol (MEG), methanol and olefins. It is projected that the annual output of CTL will reach 50 million tonnes by 2020. The current low price of petroleum is not expected to slow down the proposed CTL projects because coal prices also tend to fall as oil and gas become cheaper. Many other countries that have large coal resources are also considering coal conversion technologies.

Synthetic graphite has become a major feedstock for a wide range of premium carbon materials. Coal-derived graphene (a single atomic layer of graphite) is a major building block for the production of carbon nanotubes. While there are currently very few commercial applications of graphene, there is little doubt that the material will dominate the family of new generation materials of the 21st century, based on the extent of intensive research and development worldwide (EU, 2013; Patel, 2009; Bae et al. 2010; Aliofkhazrarei et al., 2010). Current efforts focus on methods of producing the material in commercial quantities, a key propellant for future applications (Nuwer, 2014). Potential areas of application of graphene include lightweight, strong structural composite materials, highly conductive composites, energy conversion and storage, photovoltaics, transparent screens, filtration/desalination, paints and coatings, high-temperature lubricants, semiconductors, transistors, high-conductivity supercapacitors, photodetectors, biomedical engineering and mobile and flexible/stretchable electronic devices (Palucka, 2013; EU 2013; Njoroge, 2014; Bennington-Castro, 2014; Yue, 2015; Chakrabarti, 2016; Hui, 2016). Graphene holds great promise for biomedical engineering due to its exceptional combination of properties, in particular, strength, mechanical and chemical stability, biocompatibility and conductivity. Areas of intensive studies include neurological research, orthopedics, investigative medicine and drug therapy. It has been used as a reinforcing agent to improve the mechanical properties of biodegradable polymeric nanocomposites for engineering bone tissue applications.

8.5   CONVERSION OF NIGERIAN COALS TO LIQUID FUELS

Not much research has been done on the assessment of the suitability of Nigerian coals for conversion. However, sufficient information and data on the chemical properties of the coals are available from which a useful assessment could be made. Any coal can be converted to liquid fuels but process efficiencies vary with coal rank. While the lowest rank of coal (lignite) can be liquefied, the efficiency is very low because of the high moisture content and low carbon value (Table 8.1). The yield of lignite from the drying process is only 62% compared with 93% for hard coal. Also, 60% more lignite is required than hard coal for the same product output. The actual process thermal efficiency is a function of the ($CO + H_2$) produced. The gasification technology adopted also has a significant effect on thermal efficiency. On the other hand, conversion of low-grade coals to liquid fuels is potentially attractive because of the relative global abundance and the low export potential. Also, the margin for upgrading the heat content is higher compared with bituminous coals (Figure 8.6). Furthermore, the conversion plant can be located close to coal deposits, thus eliminating the need for uneconomic transportation over long distances.

No report of any liquefaction tests on Nigerian coals could be found in existing literature but since they are all low-rank coals, conversion efficiency will be low and, considering that the country has substantial petroleum and gas resources, the economics of coal liquefaction may not be attractive. However, the low sulphur and ash contents of most of the coals are potential assets in coal liquefaction technology.

8.6   CONVERSION OF NIGERIAN COALS TO CHEMICALS

Prior to the 1950s, most of the important industrial chemicals were derived from by-product gases of coal carbonization and combustion, but, with the discovery of petroleum deposits in many parts of the world attention shifted to this source as cheaper feedstock for the production of chemicals. However, the relative abundance and wide geographical distribution of coal compared with petroleum has shifted attention back to coal and, once again, the resource is emerging as a major source of chemicals, particularly in countries like China and other countries which have abundant coal resources.

Table 8.1   Product yields from coals of different ranks. *(Tabak, S. et al., 2008).*

Variable	Lignite	Hard coal*
Coal feed as received (t/h)	913	378
Dried coal to gasifier (t/h)	570	353
Gasoline product (bbl/d)	18,000	18.000
LPG product (bbl/d)	3,300	3,300
* Low moisture, medium volatile, high ash content		

The technologies for conversion of coal to valuable syngas, hydrogen and a wide range of chemicals have been discussed above. Virtually all coals can be converted to chemicals but the process efficiency varies widely depending on the rank and chemical composition of the coal. While high-rank anthracite coal is very good for conversion to acetylene because of the high carbon content, low-rank coals are preferred because of the relative abundance and low prices. Furthermore, low rank coals tend to contain high volatile matter which makes them more suitable for the production of a wider range of chemicals, including ammonia-derived fertilizers and many tar-derived chemicals. Although low-sulphur, low-rank coals are considered the best feedstock for conversion to premium chemicals, even high-sulphur coals can be processed because the sulphur can be recovered and processed into valuable chemicals as well.

Conversion of coal to gas is always the starting point in coal-to-chemicals plants. Although some chemicals are extracted from coke oven by-product gas, most modern coal-to-chemicals plants feature a coal gasifier as the starting process. Coal gasifiers may be deployed specifically for electric power generation, or in a flexible configuration which allows the syngas produced to be utilized in several ways (Figure 8.10). Nigerian coals have not been investigated in any depth specifically to determine their potential as feedstock for the production of chemicals. However, there is sufficient data from various investigations to form a reliable basis for drawing useful conclusions about their suitability for conversion to valuable chemicals. Most of the coals have low sulphur and ash contents and all of them are high-volatile coals, hence they meet the two basic requirements. Also, the by-product gases from carbonization tests have been analyzed and found to be potentially rich in chemicals.

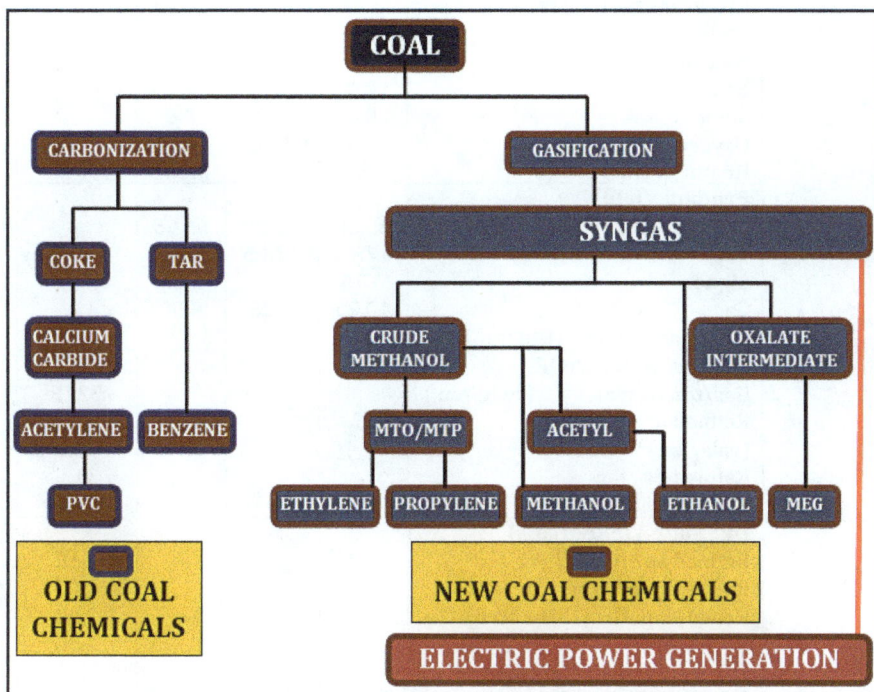

Figure 8.10   A hybrid coal-to-chemicals and power plant.

The chemical properties of Nigerian coals indicate that they are rich in chemicals, particularly Enugu and Okaba coals. Many valuable chemicals – creosote, a good wood preservative, ammonia, benzene, coal gas, coal tar, carbo-pitch etc. can be sourced from these coals. The earliest work on the chemistry of Enugu coal and its potential as feedstock for the production of chemicals was carried out between 1928 and 1937 at the H. M. Fuel Research Station, U.K. (Powel Duffryn, 1947, 1949). The coal was carbonized at 600°C in Gray-King Assay and the yields are presented in Table 8.2. For every ton of coal carbonized, the yield of the major chemicals were 9.7 litres of phenols and cresols, 29.5 litres of light oils, 59.5 litres of medium or heavy oils, and 0.442 tons of pitch. In later work by the same establishment (1947-49) on larger samples, the coal was carbonized at a temperature of 650-700°C and the tar condensed in a water trap. The tar which constituted 11.7% of the feed coal was processed in a fractionation column between 160-390°C into seven fractions and each fraction was analyzed to determine the chemical constituents. The results are summarized in Table 8.3.

Table 8.2 Gray-King Assay of Enugu coal (600°C).
(Compiled from unpublished data by Duffryn, 1949).

Thermo/Chemical Properties	1928	1934	1937	1946
Proximate Analysis				
Moisture (wt. %)	5.9	7.1	7.9	
Volatile Matter	42.8	41.1	39.8	
Fixed Carbon	42.7	41.5	43.0	
Ash	8.6	10.3	9.3	
Volatile Matter (dry, ash-free)	50.1	49.7	48.1	
Proximate Analysis				
Carbon	81.4	80.6	79.7	
Hydrogen	6.1	6.1	6.2	
Nitrogen	2.1	1.8	2.1	
Sulphur	0.9	1.0	0.8	
Oxygen	9.5	10.5	11.2	
Heating Value(Btu/lb)	14520	14550	14370	
Product Yield				
Coke, (wt.%)	65.50	67.00	66.95	
Tar "	17.45	17.65	17.65	
Liquor "	6.75	6.25	6.05	
Gas "	10.25	9.25	8.30	
Gas Heating Value (Btu/ton)	-	-	3.17m	
Carbonization Yield				
Gas/ton of coal (cu.ft/ton.)				3275
Refined motor spirit				1.27
(gals/ton)				25.2
Refined dry tar "				31.6
Refined liquor "				1400
Distillate Yield (gals)				
Refined spirit from gas				1.18
Refined spirit from tar				1.27
Heavy Naphtha (170-230°C)				2.63
Burning oil "				2.88
Fraction (270-360°C)				7.85
Solid paraffin				11.06
Hard pitch				59.70
Tar acid				2.30

Henry Balfour, U.K. (1961) carbonized Enugu coal at medium and high temperatures and carried out a comprehensive analysis of the products. The results have been compiled from various short reports to the Nigerian Coal Corporation and are presented in Table 8.4. Between 1960 and 1964, several investigations were commissioned by the Nigerian Coal Corporation and the Eastern Nigeria Government, notably by Simon Carves of the U.K., Interconsulting of Switzerland, Eisenbau-Essen and Lurgi of Germany, and Arthur D. Little of the United States of America. Although the primary objective was to produce metallurgical grade coke from Enugu coal, the data generated by these investigations provided a basis for the assessment of the potential of the coal as a feedstock for the production of chemicals and are summarized in Table 8.5.

Table 8.3 Chemical composition of Enugu coal tar.
(Compiled from unpublished data by Duffryn, 1947-49).

Fr No.	BP °C	Yield %	Composition Chemical	Yield %
1	160-200	0.585	phenols, cresols	0.176
			olefins	0.187
			aromatics	0.082
			paraffins	0.140
2	200-280	2.250	cresols & xylenols	0.425
			tar bases	0.023
			yellow oil	0.550
			brown oil	1.125
			white oil	0.011
3	280-300	1.230	diphenylene oxide	0.445
			alkylated naphthalene	0.609
			xylenols	0.175
4	300-350	0.948	tar acids	0.082
			tar bases	0.020
			naphthalene	0.082
			alkylated naphthalenes	0.133
			brown oil	0.610
5	350-375	0.503	polymerized phenolics	0.058
			cresols and xylenols	0.006
			bases	0.006
			naphthalene	0.047
			alkylated naphthalenes	0.082
			aromatics	0.302
6	375-385	0.094	red-brown resin	0.006
			colourless crystals	0.013
			yellow oil	0.025
			brown oil	0.050
7	385-390	0.503	tar acids	0.023
			white crystals-parafins	0.082
			brown oil	0.398
Residue			Pitch (softening point 76°C)	3.91

Table 8.4 Carbonization of Enugu coal. *(Henry Balfour, 1961. Compiled from the records of the Nigerian Coal Corporation)*

Product	550°C	900°C
Gas (cu.ft./ton of coal)	5050	11547
Coke (cwts)	12.84	11.5
Tar (gallons)	34.0	12.6
Liquor (gallons)	21.5	35.7
Analysis of Dry Tar		
Specific Gravity(15.5/15.5°C)	1.005	1.244
Ash (%)	0.51	0.45
Free Carbon (%)	0.79	17.67
Quinoline Insoluble	0.48	3.96
	0.130	0.450
Analysis of Pitch		
Free Carbon (%)	17.49	32.68
Quinoline insoluble (%)	0.70	4.93
Coking Value	-	57.36
Specific Gravity(15.5/15.5°C)	-	1.27
	850	870

Tar Analysis

Fraction	Volume (%)	Sp.Gr.	Tar Acids	Tar Bases	Naphthalene
Low-Temp. Tar					
Amb-180	4.19	0.856	13.4	0.62	1.3
180-205	5.82	0.920	40.0	1.86	16.7
205-235	11.11	0.925	32.0	1.22	Nil
235-290	16.24	0.932	5.0	-	-
290-360	20.86	0.964	-	-	-
260-Pitch	10.27	-	-	-	-
Pitch	28.87	-	-	-	-
Loss	2.64	-	-	-	-
High-Temp. Tar					
Amb-180	3.07	0.912	Nil	0.77	-
180-205	4.81	1.000	18.5	1.77	29.9
205-235	7.80	1.031	12.0	1.31	46.3
235-290	5.65	1.029	3.0	Nil	10.9
290-360	2.02	1.067	-	-	-
260-Pitch	4.85	-	-	-	-
Pitch	68.84	-	-	-	-
Loss	2.96	-	-	-	-

Table 8.5 Chemical properties of Enugu coal. *(Compiled from reports of various consultants from the records of the Nigerian Coal Corporation).*

Product	Simon Carves	Lurgi (520°C)	Eisen-bau Eisen (550°C)	Interconsulting				Arthur D. Little			Typical for HVB Coal	
				500 (°C)	600 (°C)	700 (°C)	980 (°C)	550 (°C)	700 (°C)	1000 (°C)	450-470 (°C)	900 (°C)
Tar, water-free (%)	6.980	17.9	11.0	12.8	8.6	7.5	8.0	13.9	9.2	9.9	6.8	4.5
Light oil (%)	6.440			0.7	1.5	1.	1.5				0.7	6.8
Liquor (%)	6.980	6.0									5.0	1.1
Gas (wt%)	12493*	7.4	13.8	5.7	11.2	17.4	19.7				11.2	18.4
H_2 in gas (vol. %)				14.2	20.1	30.6	52.7					
CH_4 " "				55.6	49.6	40.1	27.2					
CO_2 " "				11.2	7.2	5.7	4.0					
N_2 " "				3.6	2.8	2.3	2.4					
C_nH_m " "				8.5	13.2	13.0	3.6					
Calorific Value of gas (KJ/m^3)			1330	1826	1964	1818	1249	1286	1286	1286		
Coke/Semi coke (%)	61.24	68.7	75.2	74.6	70.3	67.3	66.0				64.3	62.5
* Cu.Ft.,												

There are inconsistencies in the available data presented in Tables 8.2 to 8.5 but this is understandable considering the fact that the investigations spanned a period of about fifty years. Furthermore, the sampling methods and scale of tests varied considerably. Coal is a very heterogeneous material and improper sampling can introduce significant errors in test results. The quantities and chemical nature of carbonization products also depend on the temperature as well as the method of carbonization. Since the tests which generated the results were carried out at different temperatures using different equipment and methods, the results cannot be compared directly. However, the data consistently show that Enugu coal is rich in chemicals, in particular, chemicals-rich tar, the yield of which is significantly higher than is typical of high-volatile bituminous coal. The yield of gas is also very high.

All the investigations discussed so far focused on Enugu coal but other more recent tests included other coal deposits (Afonja, 1975, Estel 1976, ROE, 1982). The results are summarized in Table 8.6 and all show that Enugu, Okaba and Orukpa coals are rich in tar which is a source of hundreds of chemicals, as well as gas which is valuable as fuel as well as feedstock for chemicals. There is some consistency in the results due apparently to the more rigorous sampling adopted in the tests. Carbonization was carried out in Jenkner-type steel retorts and the carbonization conditions were similar. The results give a good indication of yields that could be expected in commercial-scale carbonization. Analysis of the ammoniacal liquor condensate from carbonization of Nigerian coals showed that the ammonia content is very low, less than 1% (Table 8.7). However, the phenol content is very high, nearly 8%. It is unlikely that either can be converted to chemicals economically but both compounds are considered harmful to the environment and may have to be removed before discarding the liquor. The tar contents of Nigerian coals range from 9% to 14%, the highest values being for Enugu coal. Fractional distillation of the tar yielded about 36% of light oil, 15% middle oil and 12% heavy oil.

The pitch residue was around 37%. The high softening point of the residue pitch of around 93°C makes it potentially suitable for road construction in the country's tropical environment. Chromatographic analysis of the liquor fractions identified about forty important industrial chemicals that are present in significant proportions, including benzole, toluol, xylol, phenol, anthracene, naphthanol. Analysis of the crude benzole showed that it contained about 25% aliphatics (C_1-C_5) and 75% aromatics (Estel, 1976).

It should be noted that all the results presented in Tables 8.2 to 8.6 were obtained using externally heated laboratory equipment capable of batch-carbonization of around 1kg and do not necessarily correlate with what obtains under industrial conditions. For example, liquor and tar obtained from externally heated laboratory retorts differ significantly from those obtained from internally heated industrial fluidized-beds in terms of quantities and chemical composition, and products of fluidized-beds are usually more polluted by fly ash. Industrial units usually produce more tar and the low-boiling fractions, particularly light and middle oils. Most modern coal-to-chemicals plants feature fluidized-bed carbonizers and gasifiers.

Table 8.6   Chemical resource potential of some Nigerian coals.

Product	Afonja (1975)			Estel (1976)			ROE (1982)	
	Enugu coal (900°C)	Orukpa coal (900°C)	Okaba coal (900°C)	Enugu coal (870°C)	Orukpa coal (870°C)	Okaba coal (870°C)	Enugu coal (1000°C)	Okaba coal (600°C)
Tar (%,daf)	13.4	12.1	9.6	14.1	11,6	9.1	12.8	10.3
Gas (l/kg, daf)	284	290	312	280	296	309	280	255
Crude benzene (%,daf)	1.7	1.9	1.7	2.0	1.8	1.6	1.73	1.68
Ammonia (%,daf)	0.6	0.5	0.5	0.7	0.4	0.4	0.30	0.32
Chemical water (%,daf)	6.8	7.7	7.1	5.6	6.7	8.0	6.5	6.7
Coke (%db)	62.2	60.0	60.8	61.5	58.3	61.2	60.6	59.5
db = dry basis, daf = dry, ash-free								

Table 8.7   Fractional distillation of ammoniacal liquor from Enugu coal. *(Compiled from Afonja, 1975, Estel, 1976).*

TEMPERATURE (°C)	DISTILLATE	CONTENT (%)
100-180	Light oil	31-32.5.
180-270	Middle oil	14-16.5
270-335	Heavy oil	11.5-12.7
	Pitch residue	36-37.6
PITCH QUALITY		
Softening point (°C)		92-93.5

8.7 CONVERSION OF NIGERIAN COALS TO PREMIUM MATERIALS

Nigeria has two petrochemical plants and a third is under construction, hence a coal-to-chemicals plant may not be competitive. However, the flue gases from the planned coal-fired power plants can be processed into valuable primary feedstock for a wide range of prime chemicals and materials such as dyes, organic chemicals, corrosion protection coatings, graphite crucibles and electrodes, and carbon black for the tyre, paint and printing industries. This will be discussed further in Chapter 9.

REFERENCES

Afonja, A. A. (1975). "Chemical, Petrographic and Coking Studies of Enugu Coal." Nigerian Mining and Geology, 1975, 12, p. 40.

Afonja A. A. (2017). *Basic Coal Science and Technology*. SineliBooks.

Aliofkhazraei, M., Ali, N., Milne, W., Ozkan, C., Mitura, S., and J. Gervasoni (Eds) (2010) *Graphene Science Handbook: Applications and Industrialization*. CRC Press.

Asbury Carbons, (2006). *An Introduction to Synthetic Graphite.* https://asbury.com/pdf/SyntheticGraphitePartI.pdf

Bae, S., Kim, H., and Y. Lee (2010). "Roll-to-roll production of 30-inch graphene films for transparent electrodes." Nat. Nanotechnol. 5:574-8.

Bennington-Castro, J. (2014). "Future stretchable electronic devices." *Materials Research Society*, October, 2014.

Chakrabarti, A. and NS Hosmane (2016). "Crystalline graphene and method of making crystalline graphene." US Patent 9,340,430,2016.

CHINA DAILY (2016) "Significant coal-to-liquid project in production in Ningxia." www.chinadaily.com.cn/china/2016-12/28/

DOE, (2015). "Direct Coal Liquefaction." US Department of Energy. www.netl.doe.gov.

Eastman (2005). "Eastman Gasification Overview." www.eastman.com

Estel (1976). "Investigations into possibilities of preparing and carbonizing Nigerian coals from Enugu District." Estel Exploration und Bergbau GmbH, Dusseldorf.

European Commission (2013). *Graphene and Human Brain Project*. Press Release, January 28, 2013.

ExxonMobil (2014) "Methanol To Gasoline (MTG) Technology: An Alternative for Liquid Fuel Production. http://cdn.exxonmobil.com/

Gabrielsen, P. (2013). "Thinnest graphene sheets react strongly with hydrogen atoms; thicker sheets are relatively unaffected.) *Stanford Report*, January 31, 2013.

Henry Balfour, U.K. (1961). *Carbonization of Nigerian coals*. Report submitted to Nigerian Coal Corporation.

Lee, C., Wei, X., Kysar, J., and J. Hone (2008). "Measurement of the Elastic Properties and Intrinsic Strength of Monolayer Graphene." *Science* 18 Jul 2008: Vol. 321, Issue 5887, pp. 385-388.

Li, Hui et al. (2016). "Spontaneous protein adsorption on graphene oxide nanosheets allowing efficient intercellular vaccine protein delivery." *Appl. Mater. Interfaces*, 2016, *8* (2), pp 1147–1155.

Minchener, A. J. (2016). "Gasification based Coal-to-oil, gas and chemicals in China: economic and environmental challenges." IEA Clean Coal Centre www.iea.org.

Mochida, I, Okuma, O. and S. Yoon (2014). "Chemicals from direct coal liquefaction." Chem. Reviev 2014, 114(3), pp. 1637-1672.

Mukhopadhyay, R. and R. Gupta (Eds) (2013). *Graphite, Graphene and their Polymer Nanocomposites.CRC Press.*

Njoroge, J. (2014). "High conductivity supercapacitors achieved with graphene nanocomposites." *Materials Research Society*, July, 2014.

Nuwer, R. (2014). "A new method for mass producing high quality graphene." *Materials Research Society*, April 2014.

Palucka, T., (2013). "Honeycomb-like 3D Graphene Excels as a Counter-electrode Catalyst in Solar Cells." Materials Research Society Published: 27 August, 2013.

Palucka, T., (2013). "Graphene-coated Porous Silicon Opens Path to Integrated Energy Storage." Materials Research Society | Published: 15 November, 2013.

Patel, P. (2009). "Bigger, Stretchier Graphene". *MIT Technology Review,* January 15, 2009.

Powel Duffryn Technical Services Ltd., (1947, 1949). *The chemical examination and utilization of Nigerian coals and lignites.* 2nd Report to the Under Secretary of State for the Colonies, London, 1949.

ROE GMBH (1982). "Feasibility Study of the Carbonization of Okaba Coal." Final Report prepared for the Nigerian Coal Corporation. Berlin, West Germany.

Shui, H. , Cai, Z., and C. Xu (2010). "Recent Advances in Direct Coal Liquefaction." *Energies,* 2010, 3(2), 155-170; goi:10.3390/en3020155.

Tabak, S. Zhao, X., Brandl, A. and M. Heinritz-Adrian (2008). "An alternative route for coal to liquid fuel – ExxonMobil Methanol-to-gasoline (MTG) Process." *Proc., First World coal-to-Liquids Conference,* April, 2008, Paris.

Tennant, J. B. (2014). *Overview of Coal and Coal Biomass to Liquids (C&CBTL) Program.* U.S. Department of Energy National Energy Technology Laboratory. www.netl.doe.gov.

Vasireddy, S. et al. (2011). "Clean liquid fuels from direct coal liquefaction: chemistry, catalysis, technological status and challenges." Energy & Environmental Science, Issue 2011.

Yue, H. et al. (2015). "Exploration of graphene oxide as an intelligent platform for cancer vaccines." Nanoscale, 2015, 7, 19949. pubs.rsc.org.

9   Utilization potentials of Nigerian coals

9.1   INTRODUCTION

Nigeria is very rich in energy resources – coal, oil, gas, hydropower, biomass, solar and uranium. However, until recently, there had been no coherent, holistic policy on energy utilization. Existing policies had been those addressing specific energy sectors such as electricity, oil, gas, and solid minerals. The existing policies on energy had been developed from the limited perspectives of each of the sub-sectors and had resulted in conflicting policies and programmes which have not made any major impact on the energy sector of the Nigerian economy (Lukman, 2003). After about two decades of efforts to develop a coordinated, integrated, comprehensive and coherent national energy policy for Nigeria, a document was approved by the Federal Government in 2003. The highlights of the policy are summarized in Table 9.1.

Coal played a prime role in the national energy mix from 1906 when the first deposit was discovered in Enugu area, reaching its peak in 1959/60 when it supplied over 70% of the national commercial energy requirements. In 2001, the contribution of coal was only 0.02% compared with 31.9% for oil, 61.9% for natural gas and about 6.2% for hydropower (National Energy Policy, 2003). The National Energy Policy reads very much like the four National Plans of the 1970s and 1980s which listed desirable goals with no specifics or instruments for achieving the objectives. For over a decade since its adoption, there was no credible, logical move to implement the policy statements and the nation's energy mix remains unchanged.

The policy statements, objectives and strategies specific to coal are presented in Table 9.2. The Policy specifically addressed the demise of coal and proposed strategies for resuscitation. Only the third objective "To attract increased investment into, and promote indigenous participation in the coal industry" has been addressed since some of the coal deposits have attracted investment from the private sector. The nation's coal resources have been divided into ten blocks, several have been privatized and the others are being processed (see Chapter 2). Intensive exploration and quality evaluation activities are ongoing in most of the sites. Unfortunately, the Mining Act of 2007 and privatization instruments were designed to dispose of governments long list of moribund projects of which the coal Industry is one. This is evident in the statement on the mining Act of 2007 by the Minister of Mines and Steel Development in 2014: "The intention of Government is to allow the Ministry of

Mines and Steel Development (MMSD) to focus solely on its role   as policy administrator/regulator, while the private sector will act as owner-operator so as to facilitate rapid growth of the mineral sector in the economy." There were no specific clauses to enforce diversification of the nation's energy mix or local use of coal. In effect, a new owner could decide simply to mine coal and export to other countries. Fortunately, all the new owners of the privatized deposits are involved in exploration and production activities within their blocks and all of them have plans for coal-fired power plants. Also, in response to the national crisis in the power generation sector in 2014 caused by incessant disruption of gas supply to generating plants, the Minister of Power made a statement that, henceforth, potential private sector bids for concession in the coal sector must include a proposal for a captive coal-based power generation plant.

9.2   ELECTRIC POWER GENERATION IN NIGERIA

Energy, in particular, electric power is a primary indicator of the level of human and technological development of a nation. Access to energy is a primary variable in the United Nations' rating of development (Human Development Index or HDI). In 2014. Nigeria featured in the lowest 20% (154th) of the 184 countries rated, behind Cameroon, Ghana, Kenya, Equatorial Guinea, Congo, Angola (UNDP, 2014).

Table 9.1 National Energy Policy, Nigeria. *(Energy Commission of Nigeria, April, 2003).*

i	To ensure the development of the nation's energy resources, with diversified energy resources option, for the achievement of national energy security and an efficient energy delivery system with an optimal energy resource mix.
ii	To guarantee increased contribution of energy productive activities to national income.
iii	To guarantee adequate, reliable and sustainable supply of energy at appropriate costs and in an environmentally friendly manner, to the various sectors of the economy, for national development.
iv	To guarantee an efficient and cost effective consumption pattern of energy resources.
v	To accelerate the process of acquisition and diffusion of technology and managerial expertise in the energy sector and indigenous participation in energy sector industries, for stability and self-reliance.
vi	To promote increased investments and development of the energy sector industries with substantial private sector participation.
vii	To ensure a comprehensive, integrated and well informed energy sector plans and programmes for effective development.
viii	To foster international co-operation in energy trade and projects development in both the African region and the world at large.
ix	To successfully use the nation's abundant energy resources to promote international co-operation.

Table 9.2   Strategies for reactivating the coal industry in Nigeria.
(Energy Commission of Nigeria, April, 2003).

Policies	
i	The nation shall pursue vigorously a comprehensive programme of resuscitation of the coal industry
ii	Extensive exploration activities to maintain a high level of coal reserves shall be carried out
iii	Private sector as well as indigenous participation in the coal industry shall be actively promoted
iv	The exploration and utilization of the coal reserves shall be done in an environmentally acceptable manner
Objectives	
i	To promote production of coal for export
ii	To promote effective utilization of coal for complementing the nation's energy needs and as industrial feedstock
iii	To attract increased investment into, and promote indigenous participation in the coal industry.
iv	To utilize coal in meeting the critical national need of providing a viable alternative to fuelwood in order to conserve our forests.
v	To minimize environmental pollution arising from the utilization of coal
Strategies	
i	Intensifying the drive for coal exploration and production activities.
ii	Providing adequate incentives to indigenous and foreign entrepreneurs so as to attract investments in coal exploration and production.
iii	Providing adequate incentives for the large scale production of coal stoves at affordable prices.
iv	Providing adequate incentives to indigenous and foreign entrepreneurs for the establishment of coal-based industries.
v	Developing adequate infrastructure for handling and transportation of coal within and out of the country.
vi	Organizing awareness programmes for the use of smokeless coal briquettes as an alternative to fuelwood.
vii	Encouraging R & D in the production, processing and utilization of coal.
viii	Introducing clean coal technologies into coal utilization.
ix	Re-introducing the use of coal for power generation.

The International Energy Agency has proposed an Energy Development Index (EDI) on a scale of 0 - 1.0 for the assessment of access to energy, in particular, electricity and clean cooking facilities in countries (IEA, 2012). There is a strong correlation between HDI and EDI (Figure 9.1). Nigeria's rating was 0.04 compared with Ghana (0.06), Gabon (0.12). Only 45% of Nigerians had access to electricity in 2010 compared with Ghana (72%), Botswana (66%), Egypt (100%). Also, Nigeria was one of the ten countries of the world which collectively accounted for nearly two-thirds of those deprived of electricity (Figure 9.2).

The two most common indicators are generating capacity/capita and generating capacity/GDP (Figures 9.3 and 9.4). Economic and industrial development, education, healthcare, water supply, poverty reduction, good communication systems, all depend critically on availability, adequacy and reliability of electric power supply.

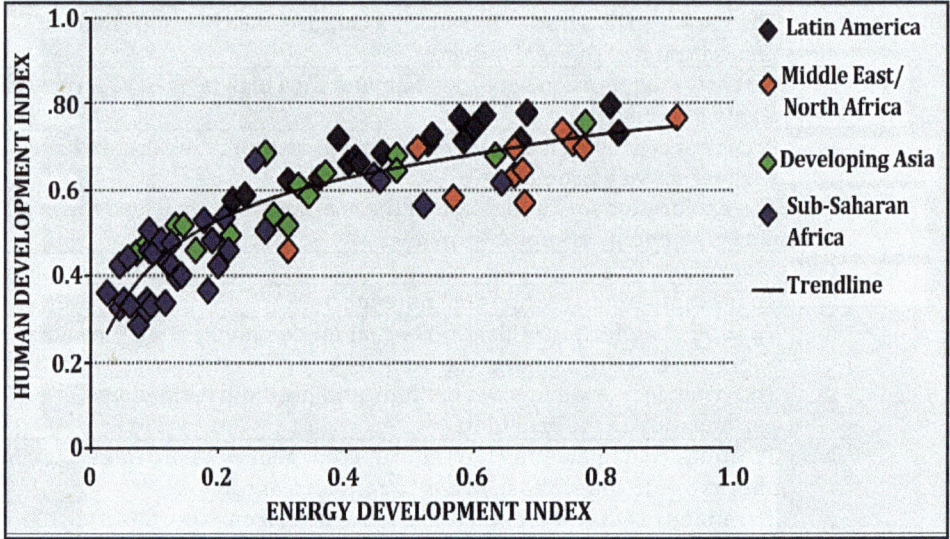

Figure 9.1 Comparison between the new Energy Development Index and the Human Development Index in 2010. *(OECD/IEA , 2012).*

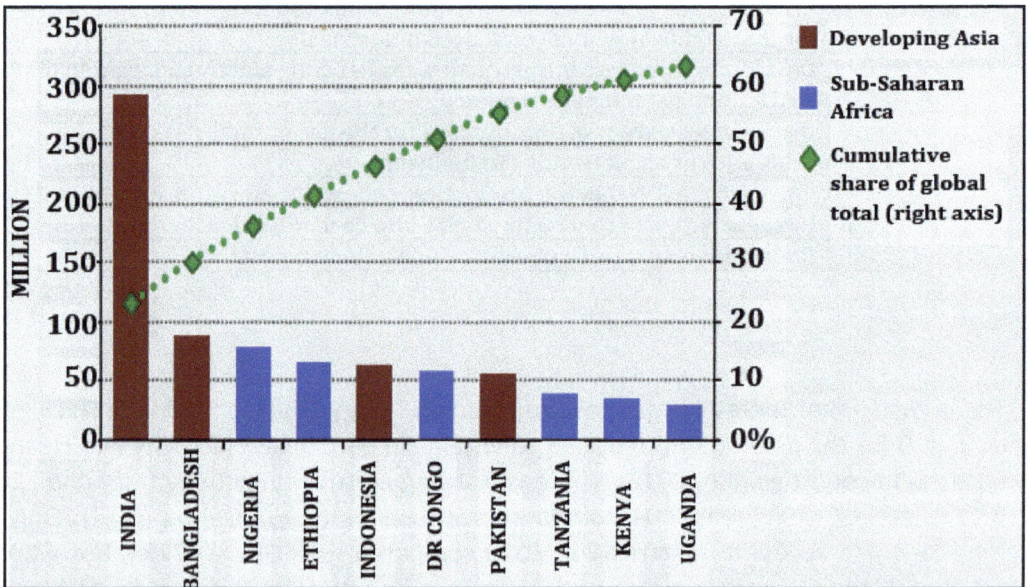

Figure 9.2 Top ten countries with the largest population without electricity in 2010. *(OECD/IEA, 2012).*

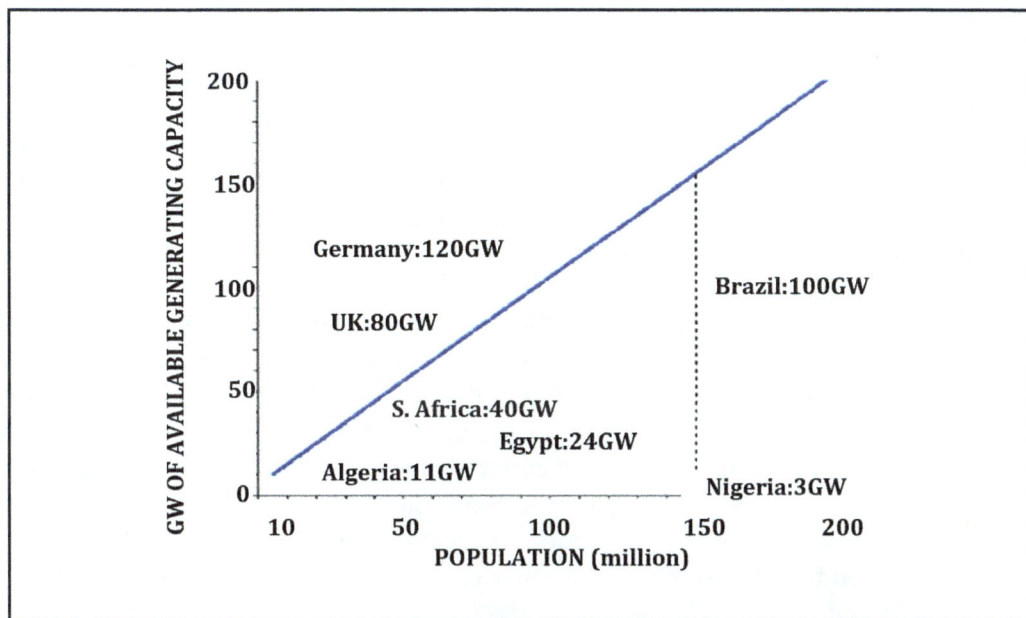

Figure 9.3 Grid-based generating capacity per capita of some countries. *(Roadmap, 2010).*

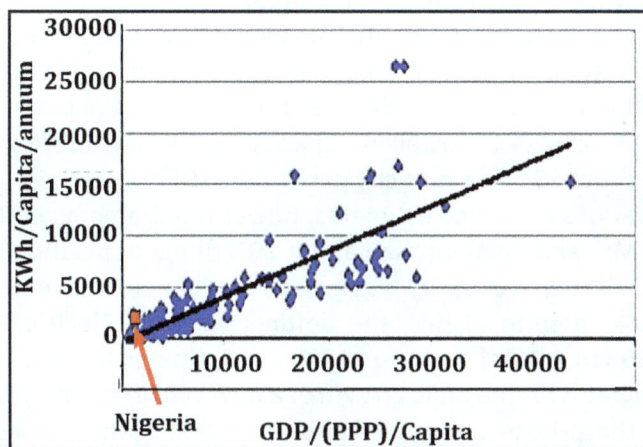

Figure 9.4 Electricity consumption and GDP in 134 countries of the world. *(Roadmap, 2010).*

Installed generating capacity is often not indicative of available electric power because poor management, maintenance and high energy losses can make a significant difference between generating capacity and useful power available to consumers. In many developing countries, up to 50% of installed capacity is often idle due to inefficient operating strategies and poor maintenance culture.

9.2.1   Historical background of the power industry in Nigeria

The first electric power generating plant was installed in Nigeria in 1896, primarily to serve the colonial masters. The plant operated on imported coal but switched to Enugu coal mine which was established in 1916 following the discovery of coal in the area in 1909. In 1929, the Nigerian Electricity Supply Company was established to provide power for the growing industrial sector. A small 6.6MW power plant was built in Calabar in 1934 and two more plants built in Oji (10MW) and Ijora (60MW) were commissioned in 1955 and 1956 respectively. The Oji plant was later upgraded to 30MW.

The discovery of oil in Nigeria in 1956 and production from 1958 marked the beginning of the demise of coal utilization in Nigeria. The first refinery was built in Port Harcourt in 1965 and three more were built in Warri (1978), Kaduna, (1980) and Port Harcourt II (1989). Low pour-point fuel oil (LPPFO), one of the by-products of the refineries became readily available and the relatively low price combined with the inefficiency of the Nigerian Coal Corporation (NCC) provided the right incentive for the conversion of the Oji power generating plant from coal to oil. The Nigerian Railway Corporation which was then the biggest customer of the NCC also converted to diesel traction and other consumers including cement plants also moved from coal to oil.

The first gas-fired (gas turbine) power was installed in Afam in 1962-63 and upgraded twice to a capacity of 776MW over the next twenty years. Another gas-fired (steam turbine) plant with a generating capacity of 972MW was commissioned in Ughelli in 1966. The first hydropower plant with a capacity 760MW was commissioned in Kainji in 1968 and a second one came on stream in Jebba (570MW) in 1985. A third plant with a capacity of 600MW located in Shiroro was commissioned in 1990. The installed capacity for electricity generation, which is 98% owned by the Federal Government, increased by a factor of 6 over the period 1968 to 1991 and stood at 5,882MW, including about 600MW produced and fed to the national grid by some of the oil companies. No further addition to generating capacity occurred over the subsequent decade. Over the years, the availability of electric power varied from about 27% to 60% of installed capacity, while transmission and distribution losses accounted for about 28% of electricity generated. (NEPA, 2003).

Based on various published estimates, power requirements in Nigeria in the early 1990s were about 15MW and no more than about 20% of the population had access to electricity. Clearly the total installed capacity was grossly inadequate and actual power availability was only about 20%. In spite of this, the defunct National Electric Power Authority (NEPA) projected a growth rate of 6-7% for 1990-2010. The crisis situation in the power sector provided the impetus for the rapid growth of a new industry: self-power providers (SPP). Most businesses in the private and informal sectors and many homes installed diesel/petrol generators. The SPP sector has grown very rapidly over the last two decades and conservative estimates put the total generating capacity in the sector at about 10MW, about three times the average power provided by NEPA.

9.2.2   Current status of the power industry in Nigeria

In spite of the impressive achievements in power generation between 1962 and 1990, power supply became progressively worse due to inefficient operation and management which caused incessant power outages (all the plants were government-owned). Supply was largely erratic and unpredictable. Inefficient transmission and distribution also caused post-generation losses of up to 30% of power produced (Table 9.3). Power supply in Nigeria per capita or per GDP is one of the lowest in the world, even in Africa (Figures 9.3 and 9.4, Table 9.4), less than 50% of the population have access to electric power, and petrol or diesel based self-generation power accounts for over 50% of power consumption. The Roadmap has set a target of 40,000MW generating capacity by 2020. As at late 2017, the generating capability was less than 7000MW. Even at the projected 2020 level, per capita capacity would still be less than a quarter of the value for South Africa (Roadmap, 2010). In an attempt to address this problem, government set up a task force on power supply in 2001. One major result of this action was the development of a National Electric Power Policy (NEPP) which led to the development of a roadmap for the privatization of the power sector of the economy.

Table 9.3   Electric Power generation in Nigeria in 2013.
(Presidency, Federal Government of Nigeria).

Electric Power Generation Capacity	2010		2013	
	Projected	Achieved	Projected	Achieved
Installed Capacity (MW)	6539	6539	8,664	6,953
Available Capacity (MW)	5500	4500	6,579	4,598
Actual Generation MW)	5000	2875	4,671	3,600

Table 9.4 Total primary energy and net electricity consumption by some countries in 2014. *(Compiled from Key World Energy Statistics, iea.org, 2016).*

Country	Total Primary Energy Supply (TPES) (toe*/capita)	Per capita consumption (KWh/Capita)
World	1.89	3,030
Africa	0.67	568
Nigeria**	0.76	144
Botswana	1.22	1,708
Ghana	0.34	357
Algeria	1.33	1,363
Egypt	0.84	1,699
Libya	2.85	1,841
South Africa	2.72	4,240
Zambia	0.64	703
Zimbabwe	0.72	543
Brazil	1.47	2,578
UK	2.78	5,131
USA	6.94	12,962
*tonnes of oil equivalent **Self power providers (SPP) output not included, estimated to be more than double the output of the national power provider		

The Electric Power Sector Reform (EPSR) Act enacted in 2005 gave legal backing to the reform and led to the incorporation of the Power Holding Company of Nigeria (PHCN) to replace NEPA. The sector was also structured into three sub-sectors:

- National Electricity Supply Industry (NESI) is responsible for generation,
- Transmission Company of Nigeria (TCN), and
- Distribution Companies (DISCOS).

The EPSR Act also provided for an independent regulatory body: The National Electricity Regulatory Commission (NERC) to regulate and monitor the Nigerian power sector. Also, a Niger Delta Power Holding Company was set up to develop ten new generating plants in different parts of the country. Unfortunately, progress through the various stages of implementation was either stalled or substantially delayed and the three major sectors took off only in the last few years. Although work started on the ten plants now known as the National Independent Power Project (NIPP), progress was slow due to changing government policy and lack of adequate financing. A Presidential Task Force was set up in 2010 to develop a strategy for revitalizing the power sector. A comprehensive Roadmap was developed by the committee and major action plans included:

- Restructuring and privatization of NEPA
- Rehabilitation of existing generation plants and transmission lines
- Privatization of existing generating plants
- Completion and privatization of the NIPPs
- Mobilization of the private sector for involvement in power generation and distribution
- Rehabilitation and expansion of the National Grid
- Development of a national gas supply pipeline network
- Diversification of the energy mix for power generation

The four thermal generating plants owned by government have been privatized through the sale of a minimum of 51% equity while concessions were granted for the operation of the three hydropower stations. In addition, the Niger Delta Power Holding Company (NDHC) was set up in 2004 to fast track the rehabilitation and expansion of the nation's power generation, transmission and distribution infrastructure. NDHC launched the National Integrated Power Project (NIPP) in 2005 which included installation of ten power plants in different parts of the country (Table 9.5). Two of the plants feature combined-cycle gas turbines (CCGT) while the others have open-cycle gas turbines (OCGT). However, seven of the OCGTs can be re-configured to CCGTs.

Most of the NIPP plants have been commissioned and are in the process of privatization. Four other plants are at the planning stage: Mambilla hydro (2,600MW), Zungeru hydro (700MW). Gurara II hydro (300MW), and Kaduna thermal, dual-fired (200MW). In order to further promote the involvement of the private sector in power generation, about 70 licenses were issued to potential private providers, in addition to the few already in operation. The total installed generating capacity presently (2017) is about 12MW from 23 generating plants feeding into the national grid, well below the targets of 20,000MW in 2015 and 40,000 in 2020 set by the Roadmap (2010). Furthermore, the available capacity is only about 50% and, in recent months, output has dropped frequently to well below 3,000MW. The problems of power generation and supply in Nigeria are multi-dimensional. Seven of the plants are over 20 years, operate obsolete equipment, and face major maintenance problems. Most of the plants source fuel (gas or fuel oil) from an unstable part of the country and oil and gas pipelines are disrupted frequently. Water levels at the hydropower stations have been dropping. The transmission grid system which covers over 13,000km of 330/132kV lines also feature obsolete equipment, leading to frequent systems collapse, and losses are estimated at between 25 and 30% due to the very long transmission lines. Often, when the power plants can increase power output, the transmission grid is unable to cope. The consumer distribution system is grossly defective and DISCOs are having difficulty paying for power purchased from the TCN which in turn cannot pay the generating companies (Nebo, 2015).

It is estimated that Nigeria needs to generate a minimum of 46,000MW to gain sufficiency in power supply. However, current installed capacity is less than 30%. Furthermore, the power transmission grid can only support a maximum of 6,500 - 7,200MW, assuming that all units are fully functional. For many years, actual power output has hovered between 2,700 and 5,000MW, less than 40% capacity utilization at its peak.

Table 9.5 Electricity generation installed capacity in Nigeria, 2014.
(Source: National Electricity Regulation Commission and Niger Delta Power Holding Company Ltd).

POWER PLANT	PLANT TYPE	CAPACITY (MW)
Afam (I-V), Rivers State	Thermal-gas turbine	987.2
Ughelli, Delta State	Thermal-steam turbine	942
Sapele, Delta State	Thermal-steam turbine	1020*
Egbin,Ogun State	Thermal-steam turbine, dual-fired (LPFO/gas)	1,320
Kainji/Jebba, Niger State	Hydro	1,330
Shiroro, Niger State	Hydro	600
Independent Power Producers (IPP)		
Shell-Afam VI, Rivers State	Thermal-Gas	642
Agip-Okpai, Delta State	Thermal-Gas	482
AES-Barges, Ogun State	Thermal-Gas	270
National Independent Power Projects (NIPP)		
Alaoji Genco, Abia State	Thermal-Gas turbine	1,131
Benin-Ihovbor Genco, Edo State	Thermal-Gas turbine	508
Calabar Genco, Cross River	Thermal-Gas turbine	634
Egbema Genco, Imo State	Thermal-Gas turbine	381
Gbarain Genco, Bayelsa	Thermal-Gas turbine	254
Geregu Genco, Kogi State	Thermal-Gas turbine	506
Sapepe-Ogorode Genco, Delta State	Thermal-Gas turbine	508
Olorunsogo Genco Ogun State	Thermal-Gas turbine	754
Omoku Genco, Rivers State	Thermal-Gas turbine	265
Omotosho Genco, Ondo State	Thermal-Gas turbine	513
Total national generating capacity		13,047.2

Twenty of the generating plants currently feeding the national grid system are designed for gas firing although two or three of them can also operate on low-pour point fuel oil (LPPFO). The focus on gas for national power generation is evident in the National Energy Policy (2003) and Roadmap (2010, 2013) documents, both of which only mention other potential energy sources as possible future development goals. The section of the Roadmap on the promotion of hydro power and other alternative fuels states: "For the foreseeable future, Nigeria will remain heavily dependent upon gas for its power generation. Nevertheless, there is room for some limited growth in hydro power plants, coal-fired power plants, and other sources of renewable power e.g. wind and solar." (Roadmap, 2010). The over-reliance on just one or two energy sources is at variance with international norms. Most countries strive to diversify energy sources for power production because of the numerous potential problems that could disrupt access to any energy source (Table 9.6). The current ratio of oil/gas to hydropower is actually over 80% due to the NIPP plants coming on stream.

In summary, the prospects for improvement power generation or supply in Nigeria in the near future are not very bright. In spite of government efforts in almost two decades to reform and develop the power sector, installed generation capacity as at 2017 is only double the level in 1991. Power output has been sporadic, oscillating between 2,500 and 4,500MW. The River Niger has been dammed upstream by other countries and the frequent problem of low level water is becoming intractable. Vandalization of oil and pipelines continues to undermine regular utilization of installed generating capacity and no sustainable solution has been found. The National Grid is obsolete and requires total redesign and modernization. Distribution companies are unable to mobilize revenue because the majority of consumers have no meters, hence they are unable to pay the Transmission Company. Clearly, there is an urgent need to diversify the primary energy base for power generation to include coal and solar energy, and to upgrade the transmission grid.

Most of government policies on the power sector have been in response to emergencies. For example, in 2010 the focus was on natural gas as the primary energy resource but there was no credible plan for a national gas pipeline network until several of the NIPPs were completed but could not access natural gas. Also, the power supply crisis caused by incessant disruption of the gas supply caused government to bring coal-based power generation to the forefront in 2014.

Table 9.6 Electricity generation by fuel, Nigeria and world average in 2012. *(Key World Energy Statistics, 2014, iea.org).*

Energy resource	Nigerian energy mix (%)	World average energy mix (%)
Coal	-	40.4
Oil and gas	70	27.5
Hydropower	30	16.2
Nuclear	-	10.9
Other*	-	5.0
*Geothermal, solar, wind, biomass, etc		

9.3   FUTURE PROSPECTS OF COAL-BASED POWER GENERATION

Fossil fuels currently account for about 85% of the world's primary energy supply and projections indicate that the level will drop by only 5% in 2035 (Figures 9.5 & 9.6). Coal's share which is currently about 30% will drop to around 25%. About 90% of anthropogenic CO_2 emissions come from fossil fuel combustion and coal accounts for nearly half. The world economy is expected to almost double over the next two decades and much of the expected growth will be driven by emerging economies. Economic growth is a key determinant in the growth of global primary energy demand and about 80% of the growth will be in the power sector. Electricity is the fastest-growing form of end-use energy consumption and demand is expected to nearly double by 2040, most of the projected growth in demand being in emerging countries.

9.3.1   Status of global coal-based power generation

Coal is still the most important primary energy source for global electricity supply and currently accounts for 41% of global electric power generation, the main reason being its wide availability and affordability compared with other energy sources. Coal-fired plants supplied 30% of electric power in the United States in 2016 (over 60% in some states). Australia, Canada and many other countries of Europe, Asia and Africa rely on coal for 60-90% of power generation. Over 70% of global steel production relies on coal. Clearly, the prospects of finding a suitable replacement for coal in the global primary energy mix in the foreseeable future are very remote. All projections indicate that coal will remain dominant in the global primary energy mix and still provide about 30% of the world's demand for electricity in 2035-2040, in spite of international effort to reduce dependence because its high pollution propensity compared with all other primary energy sources (Figures 9.7 & 9.8).

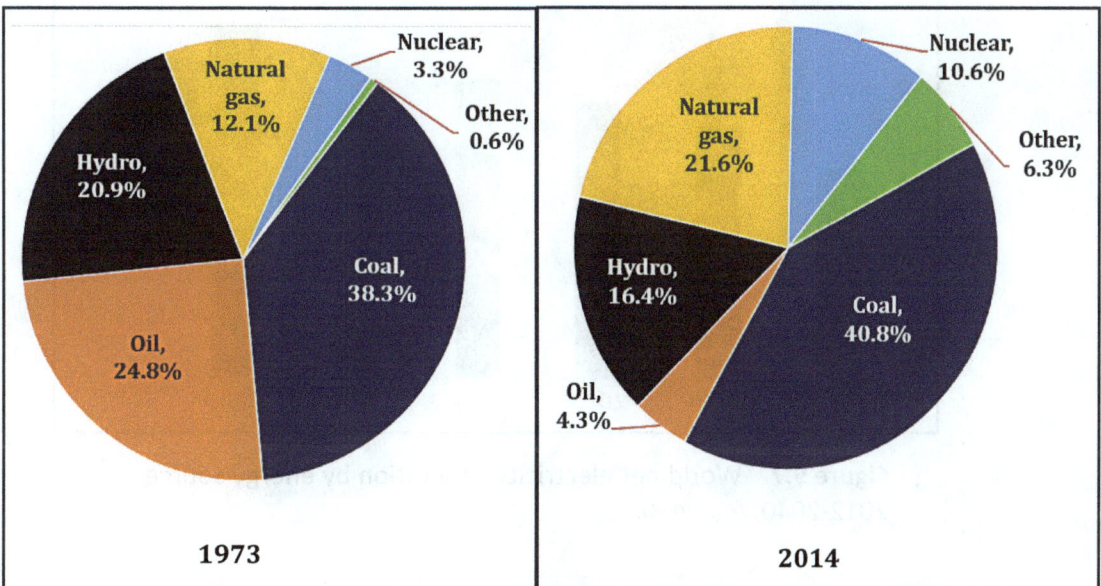

Figure 9.5   Global electric power generation by fuel, 1973 & 2014.
(International Energy Agency Key World Energy Statistics, 2016, iea.org).

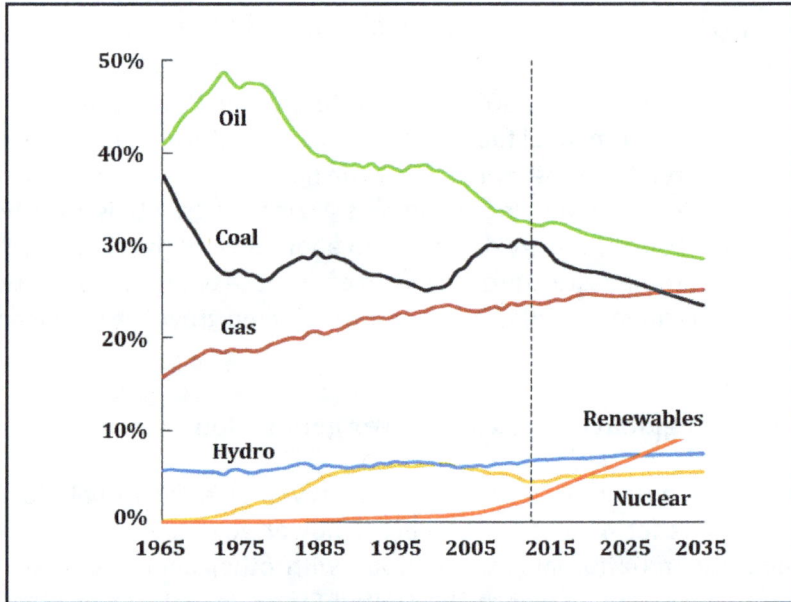

Figure 9.6   Projected share of global primary energy by fuel in 2035. *(BP 2017).*

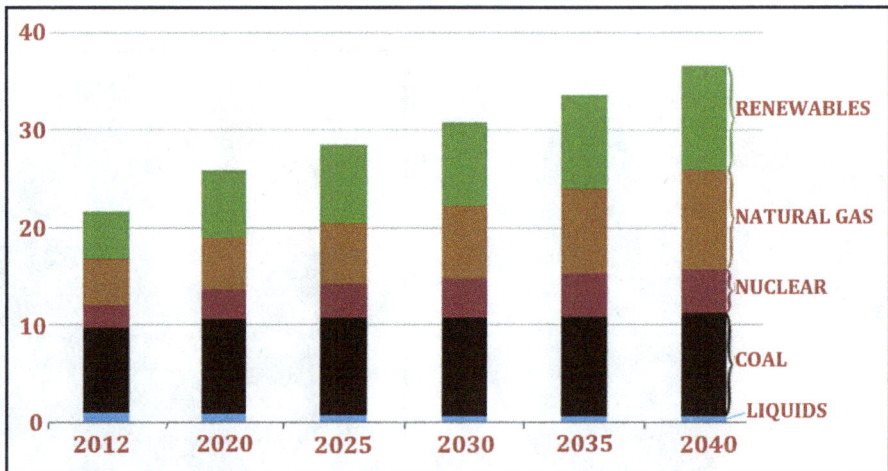

Figure 9.7   World net electricity generation by energy source 2012-2040. *(eia, 2016).*

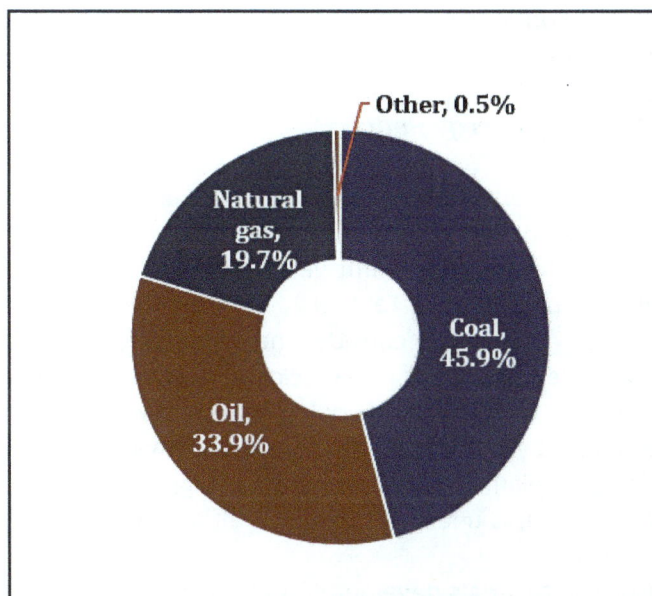

Figure 9.8 2014 fuel shares of CO_2 emissions from fuel combustion. *IEA, 2016).*

The main reason for the slow pace of phasing out coal and other fossil fuels is the fact that every other potential substitute that is more environment-friendly has other major issues. Renewables are the fastest growing fuel source, expected to quadruple in the next two decades, driven by continuing competitiveness, and renewable-based generation is expected to account for all new power generation capacity in 2040 (iea, 2016). However, the overall contribution to the global primary energy demand will still be low, only about 10% (Figure 9.6). Solar, wind and geothermal energy are site-specific and bioenergy is weather-dependent. The contribution of hydropower and nuclear power in 2040 will each be less than 10%. Apart from the fact that hydropower is weather-sensitive, resistance against construction of new dams is strong particularly in developed countries because of environmental issues. Also, several accidents in recent times involving nuclear power plants have slowed down proliferation significantly. It is now widely accepted that decarbonization of the global fuel mix will be a gradual process and the only pragmatic option in the foreseeable future is to clean up coal and other fossil fuels.

9.3.2   Status of coal-based power generation in Nigeria

In pursuance of the mono-energy policy, government and the private sector have made major investment in the development of gas distribution pipeline networks, all emanating from the Delta region. However, incessant disruptions of gas supply due to systems collapse, vandalism and frequent maintenance shutdowns have caused major problems for the generating plants. Most of the NIPP plants are either operating sporadically or unable to take off because of shortage of gas supply. The average power output in the first five months of 2017 has been less than 3,000MW. Private companies, in particular, cement plants are planning to install captive coal-fired power plants and will import coal until the privatized mines become operational. Possible options for expedited repositioning of the power sector are discussed in the following sections.

9.3.2.1   *The National Energy Policy*

Although the National Energy Policy (2003) document proposed a number of future measures for diversifying the country's energy mix, none of the options had been pursued with any dedication until the gas crisis of recent years. The clauses which specifically address coal utilization have been presented in Table 9.2. The problem of inadequate and unreliable gas supply has caused government to consider the option of diversification to coal-fired power generation in 2013. New owners of coal block are now obliged to install coal-fired generating plants at the vicinity of the mines and coal-fired plants were projected to supply 30% of the national power output by 2015. Although the decision was not retroactive, owners of blocks sold earlier had already planned for captive coal-fired power plant. It should be noted however that, as at 2017, no plant is under construction although several of the concessionaires are already active on site.

Characteristically, Nigeria's development plans are in discrete modules, with little or no consideration for a holistic systems analysis to determine the sourcing and procurement of the required inputs. A typical example is the National Integrated Power Project (NIPP) which was initiated in 2003 and involved the construction of ten gas-based power generating plants located in different parts of the country. There was no plan for the supply of gas to the various locations until construction of the plants were nearly completed, hence most of them have remained idle (NEBO, 2015). An emergency National Gas Infrastructure Masterplan has been developed and is currently being executed in phases. The fact that none of the ten NIPP plants is coal-fired is also a good example of defective planning. The five coal-fired power plants that are currently being planned are at the initiatives of the owners of the privatized coal blocks, not a deliberate government policy thrust. However, government's recent announcement that integrated coal-fired power plants would be a prerequisite for the purchase of any of the remaining coal blocks is a welcome development.

The National Energy Policy needs a comprehensive review to bring to the forefront a sustainable roadmap for diversification of the primary energy base of the country to include coal, biomass and solar energy. Some northern parts of the country have sufficient solar energy density to sustain solar power plants which could be configured as stand-alone or in a hybrid system with coal, oil or natural gas. In 2016, Morocco commissioned the first phase of what will be the world's largest concentrated solar power plant, funded by the Clean Technology Fund of the World Bank. The solar plant located in the Sahara desert uses the concentrating solar power (CSP) system which, unlike the conventional photovoltaic system, is capable of capturing solar energy and generating power on cloudy days and even after nightfall. A similar project could be suitable for the northern parts of the country.

9.3.2.2   *Distributed generation strategy*

The most important input of an electric power plant is the fuel and the optimal strategy is to domicile the plant at or near the source of fuel. The power generated is fed to the national grid for distribution. This strategy, also known as *distributed generation,* exploits the relative ease of transportation of electric power compared with fuel. It also greatly minimizes power generation problems which arise when main fuel supply routes suffer disruption. Clearly the distributed strategy option was not considered in planning the NIPP projects but the recent decision by government that new proposals for the purchase of any of the remaining coal blocks must include a domiciled coal-fired generating plant is noteworthy. Nigeria has an extensive national power transmission grid, with nodes in all the major areas of fossil fuel deposits and hydro projects hence feeding power generated in any part of the country to the grid should not be a problem (Figure 9.9).

The total planned generating capacity of the proposed coal-fired generating plants is about 4,000 - 5,000MW (Table 9.7). However, only the Zuma project has reached the design stage. Also, extensive exploration and mining activities are evident at the site (Figure 9.10). The Zuma plant in Itobe, Kogi State has a design capacity of 1,200MW to be developed in two phases of 6,00MW each, and provisions for an ultimate capacity of 3,000MW. The first phase of the Zuma power plant will feature four 150MW units to be commissioned serially between 2016 and 2018. The fluidized-bed combustors will burn sub-bituminous coal from the Okaba opencast mine, transported by truck over a distance of about 120km. However, there are plans for a rail link between the plant and the mine.

It is noteworthy that both the National Energy Policy and NERC guidelines for licensing coal-fired plants insist on clean coal technologies in order to mitigate the negative effects of coal combustion on the environment. However, it is important to monitor and enforce compliance, especially the capture and disposal of carbon dioxide which is the reference anthropogenic gas. The Zuma project has adopted circulating fluidized-bed coal combustion technology which makes it relatively easy to capture and treat most of the anthropogenic compounds in the combustion flue gases. There is no information on the proposed configuration of the other power plants. The Zuma project plan also includes production of coal-based chemicals and smokeless fuel as a substitute for charcoal, especially in the northern parts of the country.

9.3.2.3   *Future prospects of coal-based power generation in Nigeria*

Government recently set a target of coal-fired power generation for 30% of the nation's total power generation installed capacity by 2020. Five plants with a combined installed capacity of about 4,200MW are already at the planning stage and could be operating by 2020. With a total installed capacity of around 13,000MW in 2015 and ongoing projects, a generating capacity of around 20,000MW in 2020 is feasible. In effect, over 20% of total installed generating capacity could be coal-based by 2020. This would be a major achievement for the energy diversification effort, considering that coal contribution in 2016 was zero. The new government policy that all new owners of the privatized coal blocks must install domiciled coal-fired generating plants means that at least five more coal-fired plants (in addition to the five licensed already) could be in operation by 2020. This could push the contribution of coal to power generation to around 40%. It should be emphasized however that there could be major gaps between installed capacity, power generation and actual power available. Non completion or vulnerability of the national gas pipeline grid could have a major effect on power generation while an incompetent national power grid would greatly diminish the amount of power transmitted to consumers.

All the major coal deposits in Nigeria are close to the existing national power grid and planned extensions could bring them even closer (Figure 9.9). In addition, several cement companies are planning captive coal-fired plants and Government's current policy of encouraging private companies to generate more power than needed and sell the excess to the national power grid could further boost the contribution of coal to power generation. However, the obsolete national transmission grid will remain a major constraint unless current plans for modernization are expedited.

Table 9.7 Planned coal-fired electric power generating plants. *(Nigerian Energy Regulation Commission).*

Power Plant	Coal source	Capacity (MW)	Technology
Zuma Energy, Itobe, Kogi State	Okaba, Ogboyoga I & II	600/1200	Circulating fluidized-bed
HTG Pacific Energy, Nsukka	Ezimo	1000	?
ESSAR, Enugu	Enugu-Inyi	600	?
Sepco-III-Pacific Energy, Gboko	Orukpa	1200	?
Trombay Power Generation, Wajari, Gombe State	Gombe lignite	500	?
Simang Group	Enugu/Onyeama	?	?

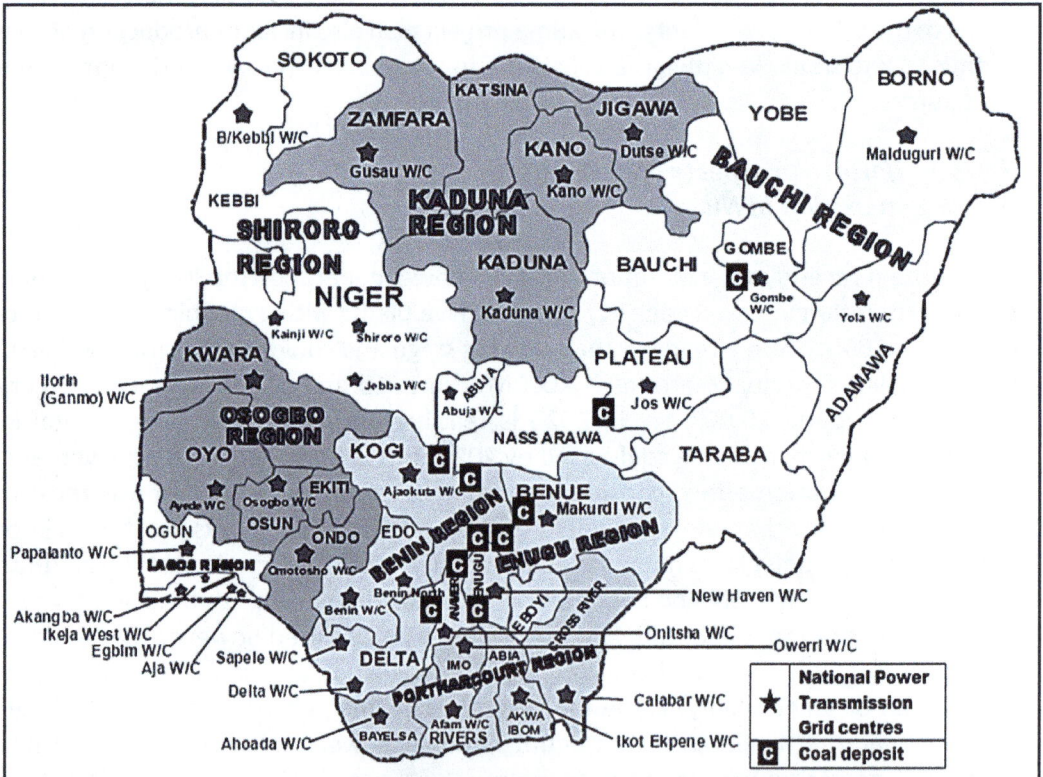

Figure 9.9   Power Transmission Grid, Nigeria. *(Adapted from Roadmap, 2010).*

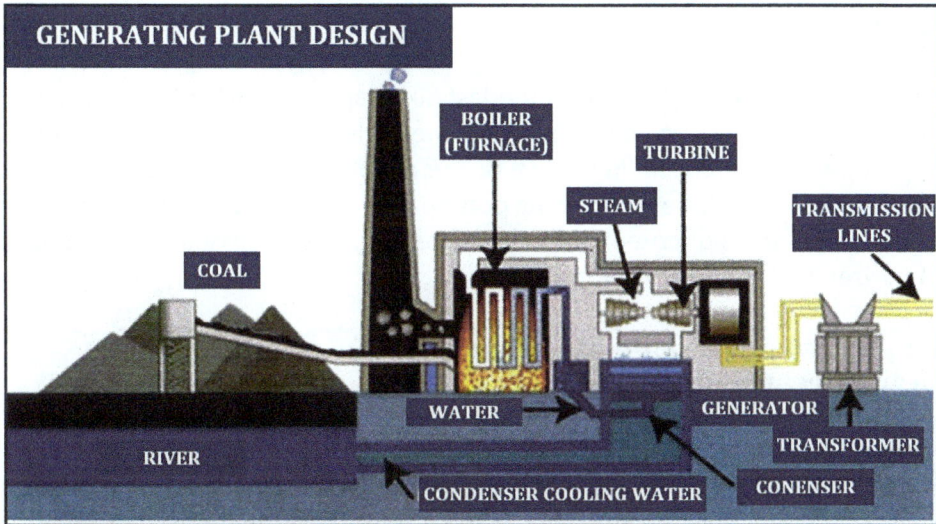

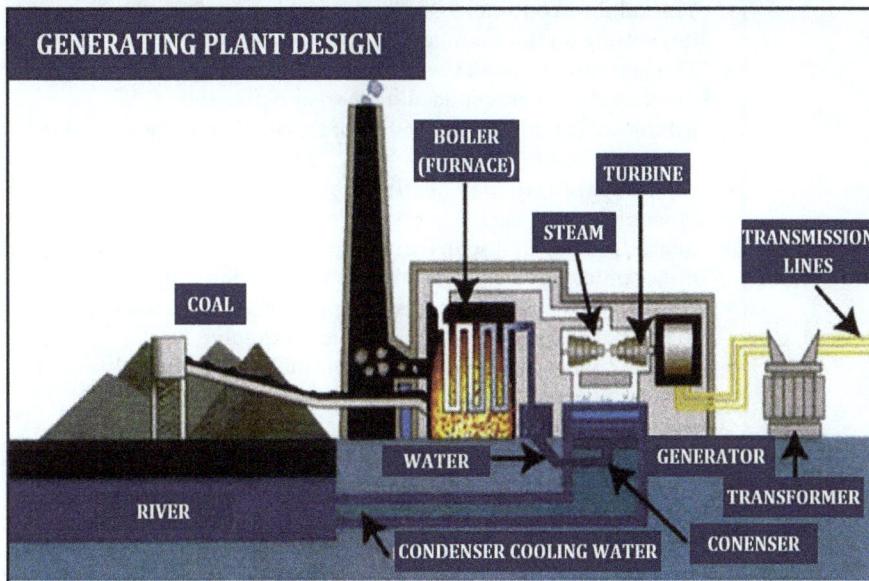

Figure 9.10   Eta-Zuma coal-to-power project, Itobe, Kogi State, Nigeria. *(Ezuma, 2014).*

9.4   POTENTIALS FOR MAXIMAL UTILIZATION OF NIGERIAN COAL RESOURCES

Information and data presented in the last six chapters on Nigerian coals have provided a useful basis for assessing the utilization potentials of the resources. The most important properties are summarized in Table 9.8. The major prospects are in the promotion of local utilization, although there is also some potential for export, particularly of smokeless fuel and chemicals. The areas of potential use and prospects are discussed in some depth below and summarized in Table 9.9.

9.4.1   Nigerian coals for blast furnace iron and steelmaking

All investigations carried out over nearly a hundred years have come to the same conclusion about the lack of caking ability of Nigerian coals. This makes them unsuitable for metallurgical coking. Numerous tests have shown that no more than about 10% of Enugu coal can be incorporated in a coking blend and the component would have to be pulverized or compacted into briquettes. The Ajaokuta Steel Plant was designed around a blast furnace system and, had it taken off, would have required about a million tonnes of coking coal annually in the first phase with a design capacity of one million tonnes of products a year, increasing by a factor of five in the final phase of 5 million tonnes.

Table 9.8 Summary of the technical quality of Nigerian coals.

• Nigeria has abundant resources of coals, mostly of sub-bituminous and lignite ranks;
• Only Obi/Lafia coal is marginally in the bituminous rank
• The sulphur contents of all the coals with the exception of Obi/Lafia coal are very low (less than 1%);
• The sulphur content of Obi/Lafia coal ranges from 3% to 11% depending on the seam and mode of sampling;
• The ash content of all the coals are within acceptable limits for most applications, especially when beneficiated;
• All the coals with the exception of Obi/Lafia coal respond well to beneficiation;
• All the coals have high heating (calorific) values;
• Most of the coals are susceptible to weathering which degrades vital characteristics, in particular, heating value, and may render them susceptible to spontaneous combustion;
• None of the coals has any caking properties, with the exception of Obi/Lafia coal which is marginally cakeable;
• All the coals with the exception of Obi/Lafia coal are potentially suitable for heating, steam-raising and power generation;
• All the coals are potentially suitable as feedstock for the production of liquid fuels and a very wide range of premium industrial chemicals.
• The prospects for exportation of any of the coals are low primarily because of the low heat value-to-transportation cost ratios, and also because of the high susceptibility to weathering during transportation and in storage.

Table 9.9 Potential areas of utilization of Nigerian coals.

AREAS OF POTENTIAL USE	PROSPECTS
Iron and steelmaking	Very low
Cement manufacture and similar industries	Very bright
Power generation	Very bright
Conversion to chemicals	Very Bright
Conversion to smokeless briquettes	Very Bright
Conversion to liquid fuels	Very low
Export	Low

The implication of the deficient quality of local coals is that about 90% of the annual coking coal requirements amounting to almost a million tonnes for Phase 1 would have to be filled by imports. One of the major constraints had been the importation of coking coal in such large quantities and transportation from any port to the remote location of the plant which is not served by any rail system or river transportation. The Niger river which has been dredged several times remains unnavigable because of continuous re-silting. Also, the rail link which was designed to link up the nearby Itakpe iron ore mine to the Warri port remains uncompleted.

Extensive tests on Nigerian coals have shown that most of them are potentially suitable for briquetting to produce formed-coke of metallurgical quality. Adoption of this process by Ajaokuta Steel Plant would have virtually eliminated or substantially reduced dependence of the plant on imported coking coal. However, the formed-coke technology which has been in existence for over fifty years has never been adopted by most steel plants worldwide. In spite of the proven comparable and in some cases, superior properties compared with traditional coke, there has been no investment in any commercial plant capable of producing enough products for the extensive commercial-scale tests needed to confirm the potential of formed-coke. The use of formed-coke has been limited to fueling foundry cupolas and incorporation in relatively small proportions in blast furnace charges. One major reason for the slow adoption of formed-coke technology has been the perennial instability in the global steel industry which has stalled investment in new plants and new technologies. Furthermore, many major steel plants either have long-term contracts with coking coal suppliers or own the mines, hence have no need to venture into new technologies. Also, blast furnace operation is more of an art than science and successful operation is based on long-term experience and operators are reluctant to embark on a new learning curve.

9.4.2   Prospects of a revitalized Nigerian steel industry

The national steel project comprised two integrated steel plants and three rolling mills. Planning for a national iron and steel industry began in 1958 and the original proposal by several European consultants was for a rolling mill designed to process scrap metal into structural steel which was then in high demand because of the numerous construction projects. With backward integration, an iron and steel making unit could be added. The project failed because government could not resolve the political issues on location of the proposed plant. Another attempt in the 1960s failed due to the political crisis which led to the civil war but was revisited immediately after the war in the early 1970s, as part of a bilateral agreement with the defunct Soviet Union.

9.4.2.1 *Ajaokuta Steel Project*

Initial planning for Ajaokuta Steel Plant began in early 1970, coordinated by the Steel Development Unit of the Federal Ministry of Industry working with Russian consultants. Extensive survey of the country was carried out to establish a sustainable raw materials supply base. The plan proposed a three-phase development, starting with a 1.2 million-tonne/year blast furnace to produce structural rolled materials. The plant would be upgraded in two phases to 5 million tonnes a year to produce manufacturing grade steel and flat products. The National Steel Development Authority (NSDA) was set up to manage the project. Integrated steel plants are usually located near a sea port or rail line, and in close proximity to a town where the workers (often in thousands) can reside and commute. The decision to locate the plant in Ajaokuta was purely political (the location was not mentioned in the feasibility study), based on the need to spread development. It was virgin land, the nearest town being over 100 kilometres away. There were no roads, rail lines or seaports anywhere near the site. Hence, the initial investment included the building of a steel town, roads, rail line to link the port in Warri, dredging of River Niger, and construction of a sea port at Baro, a few hundred kilometres from the site. Apart from building the steel town and a road to link Lokoja, none of the other projects was ever concluded. In its nearly forty years of existence Ajaokuta steel plant has never produced any molten steel even though all the core units (blast furnace, coke oven, sinter plant, oxygen converter and light section rolling mill) had been in place. Only the light section rolling mill was commissioned in 1983 and had operated intermittently over a few years, processing mostly imported billets.

9.4.2.2 *Delta Steel Project*

The National Steel Development Authority (NSDA) was established in 1971 and was expected to take over the functions of the Steel Unit of the Federal Ministry of Industry, but this did not happen. Instead the Unit began planning a rival steel project of the same size, to produce the same product mix. The Unit was working with a German consortium and decided to adopt the direct reduction process which was then at the initial stages of development. For no justifiable reason, government adopted the project and located it in Ovwian Aladja, Delta State. In effect, two parallel projects designed for the same product mix were approved by the same government. The combined product mix of the two plants was 2 million tonnes of structural steel at a time when local demand for steel products was only 600,000 tonnes, mainly flat products (see Afonja, 1979; Afonja and Ajakaiye, 1981; Oyeyinka, 1995). It should be mentioned however that, while the NSDA was unable to resolve the numerous intractable problems with the Ajaokuta Steel project, construction of Delta Steel plant began in 1977 and production started in 1982. The Midrex direct reduction unit operating on natural gas had two modules with a combined capacity of 1.2 million tonnes/year of direct reduced iron, the world's largest at the time.

The only major constraint to the DSC plant was the requirement of high grade iron ore which had to be imported. The project had a very good chance of success and all the major units were being run by Nigerians within a few years. However, the company was never allowed to run as a commercial enterprise. All the major decisions, including the pricing of products were taken by non technocrats at the Federal Ministry of Steel Development. The company never achieved more than 25% capacity utilization in over 15 years of operation because it only operated when the Ministry could provide funds for importation of iron ore and spare parts. In the meantime the company had nearly 5,000 employees (the number required for full plant capacity utilization).

In the early 1980s, it had become clear that the demand for flat products was much higher than for structural sections, particularly for the automotive industry which government was also promoting. Production of flats had been scheduled for the second phase of Ajaokuta Steel plant but the first phase had not even taken off. The Federal Ministry of Steel Development commissioned a study to determine the best way to fast track production of flat products in the country. The report suggested modification of Delta Steel Plant as the most feasible option (Afonja et al., 1983). The plant had three idle electric furnaces and continuous casting lines. All that would have been required was the modification of one of the continuous casters to produce blooms and installation of a flat products rolling mill. The proposal was rejected by government because of the need diversify investment to other parts of the country. Instead, another study was commissioned for a new flat products plant. Oku Iboku, Cross River State was selected as the site for the project but it never took off.

9.4.2.3   *The Steel Rolling Mills*

Delta Steel Plant was designed as a fully integrated unit, with two direct reduction modules feeding four electric arc furnaces which supply molten steel to four continuous casting lines. The hot billets from the casters were to be fed to four reheating furnaces and rolled on four lines of section mills. For no other reason than diversification of development, and against informed advise, Government decided to excise three of the four rolling mills and locate them in Osogbo, Jos and Katsina. The assumption then was that billets would be produced by DSC and transported to the three rolling mills. It was clear from the beginning that this would not be possible. There was no rail line in the whole of the Delta region and, because of the marshy nature of the region, road transportation was not feasible. This decision marked the beginning of the end of all four projects. Delta Steel Company was left with three idle lines while none of the three rolling mills had billets to roll. The mills eventually resorted to importation of billets but operation was epileptic. Each of the mills had a design output of about 210,000 tonnes of rolled products annually but the typical output was less than 10,000 tonnes. Furnaces had to be shut down and restarted and this led to fast deterioration of the refractory lining and frequent replacement. All refractories needed had to be imported.

9.4.2.4   *The Independent Steel Rolling Mills*

While the national steel projects struggled for survival, several foreign investors established rolling mills melting scrap and producing structural rolled products. Over the years, the number has grown to about twenty, supplying some of the country's requirement of structural steel. It is interesting to note that the first of these mini steel plants was set up by the defunct Eastern State of Nigeria in the early 1960s but failed due to government bureaucracy. Because the plants process poorly sorted scrap, the quality of products is low. All the independent rolling mills are small and the combined capacity is less than 5% of the country's annual demand for steel products and the balance is filled by importation.

9.4.2.5   *Current status of the steel projects and the way forward*

Three of the major national steel projects - Ajaokuta steel plant, Nigerian Iron Ore Mining Company (NIOMCO), Itakpe, and Delta Steel Company Aladja - were concessioned to Global Steel Holdings Limited, an Indian company in 2005 but two have been recovered due to poor performance by the new owner. Ajaokuta Steel Plant is still comatose and under a sole administrator but Delta Steel Company is now owned by Premium Steel and Mines Limited (PSML). The company is producing rolled products entirely from scrap but plans to receive iron ore from NIOMCO in the near future. The three rolling mills in Osogbo, Jos and Katsina have also been privatized and are now owned by Dangote Integrated Steel Rolling Company, Zuma Steel West Africa Limited and Dana Steel Nigeria Limited respectively. All the mills are currently processing scrap but have plans to backward-integrate to ore processing. In order to achieve this goal, all the plants need to install electric arc furnaces which require uninterrupted power supply. Zuma plans to install a coal-fired power plant on site to supply all the power requirements of the Jos plant.

Although about 98% of the infrastructure and equipment of Ajaokuta are in place, there are still many formidable problems to be resolved, notably supply of raw materials, in particular, coking coal, and rehabilitation of plant equipment which is obsolete and largely vandalized. Fortunately the core of the plant comprises the sinter plant, coke oven, blast furnace and oxygen converter all of which have no life span as long as they are re-lined with refractories as due and auxiliary equipment such as burners, blowers, motors and control equipment are updated regularly. In effect uninterrupted supply of major raw materials would be the critical determinant of successful reactivation of the plant. Once a blast furnace is fired, it must be fed with raw materials and tapped continuously day and night for about five years when relining would be due. Since nearly a million of coking coal and other auxilliary materials and spare parts will be imported annually, a rail link to a port would be critical. Construction of a rail line to link Itakpe and Ajaokuta which began in the 1980s was abandoned. However, a new contract was signed in 2017 with a Chinese consortium for a 272km line to link the Warri port, Itakpe and Ajaokuta, with an extension to link Baro and Kaduna. Also the Warri port is to be expanded. If completed successfully the project will go a long way in solving transportation of raw materials and finished products by rail to most parts of the country.

Another major obstacle to the reactivation of Ajaokuta Steel Plant is the need to find investors willing to inject the enormous funds which would be required to turn the project around. An estimated $7 billion has been spent so far, by far the most expensive plant of its size in the world, and several billion additional dollar funding would be required to complete the project. The global steel market is in depression and many steel plants have closed down in recent years. Most countries including Nigeria are fighting dumping by state-supported Chinese producers. Strict protectionist measures would be required to protect local steel industry and ensure survival.

Delta Steel was sold to GINL-GHSL in 2005, then retrieved and resold to Premium Steel Mines Limited (PSML). The current owners are producing structural steel from scrap, but have plans to reactivate all other units. Although DSC was a rival project, it has turned out to be the more suitable technology compared with Ajaokuta blast furnace technology. The direct reduced iron (DRI) technology adopted for the project was at its infancy in the 1970s when the project was commissioned, but has developed very rapidly to become the technology of choice for emerging countries. The DSC project was one of about ten gas-based Midrex DRI plants installed in developing countries around the same time but it was the biggest of them all. The plants in Brazil, Argentina, Venezuela, Egypt, Saudi Arabia, India, South Africa all achieved full capacity utilization within a few years and they are all still in operation, most of them upgraded to much larger capacities. The attributes which make the DRI process suitable for developing countries are many. The modular design affords the opportunity to start small (as low as 50,000 tonnes/annum), and scale up gradually to over a million tonnes. The minimum start-up economic size of a blast furnace is one million tonnes a year. Any module can be shut down for repair or in response to low demand for steel products. The technology is relatively simple and fuel requirement is very flexible. The DSC plant was designed for natural gas but can operate on gasified coal (syngas) or biogas. Another version of DRI technology (Hyl) operates directly on solid coal of any grade. The completion of the Ajaokuta-Kaduna rail project will place DSC at a major advantage. High-grade iron ore can be obtained from Itakpe by rail (or from Liberia by ship) and direct reduced iron or billets can be supplied by rail to three rolling mills (up to Kano for the Katsina plant). Also the products of the plant can be supplied to most parts of the country by rail. Furthermore, relatively simple modification will enable the plant to produce flat products and manufacturing grade sections, all of which are being imported currently.

9.4.3   Potential of Nigerian coals in blast furnace steelmaking

If Ajaokuta steel plant can be resuscitated about a hundred thousand tonnes of local coals will be required annually to blend with imported coking coals. The coal will probably be sourced from Okaba and Orukpa coal mines in Kogi State. It is doubtful whether the new owners of the plant would consider formed-coke technology which could utilize 80 - 100% local coals since the technology is still largely untested. However, it is conceivable that Zuma could consider adopting the NuEnergy formed-coke technology which is running very well in the United States and processing sub-bituminous coals into formed-coke of metallurgical quality. The company owns several coal blocks and has already expressed interest in producing smokeless fuel. A NuEnergy plant can be designed to produce both and formed-coke can be supplied to Ajaokuta Steel Plant by rail.

9.4.4   Nigerian coals for direct reduction steelmaking

The special features which make the DRI steel making process suitable for developing countries have been discussed in the last section. Most of the countries have no coking coal but have either gas or coal. Since coal of any grade is suitable, even countries which have neither coal nor gas can import coal cheaply from many sources. Nigeria has extensive natural gas resources mostly located in the Delta region. Gas-based DRI process is much more efficient than solid coal systems and should be the preferred option for a potential DRI project close to the national gas pipeline grid. However. Captive DRI plants could be located at or near coal mines, particularly in areas that are far from gas pipelines. Also, coal can be processed to syngas in order to take advantage of the more efficient gas-based DRI process. Direct reduced iron is a very versatile raw material for steel production and could replace scrap which is becoming increasingly scarce and does not produce good quality steel. Rolling mills that already have electric arc furnaces can switch to DRI without any problem. Furthermore, there is a good export market for direct reduced iron.

9.4.5 Nigerian coals for industrial heating

The cement industry was one of the major customers of the defunct Nigerian Coal Corporation up to the late 1960s. However, the Corporation was grossly inefficient and customers suffered from irregular deliveries. With the increasing availability of heavy oil at competitive prices, all the cement plants converted to oil. Developments in the natural gas industry in the last two decades have led to the construction of a gas pipeline grid from the Delta region to Lagos and many branches are under construction. This prompted some of the older cement plants and new ones to adopt natural gas. However, the incessant disruptions of the gas pipelines due to technical failures and vandalism have caused major disruptions to gas supply and electric power generation, and prompted some of the major cement plants in the industry to consider moving to coal either for firing the kilns or for in-plant electric power generation or both. The cement plant located in Gombe is already exploiting the nearby coal deposit and Dangote Cement Company which operates two of the biggest plants in Nigeria is installing a coal-fired power plant in one of its factories, and will import coal from South Africa because there is no functional coal mine in the country.

The prospects for coal in the cement industry are very bright, particularly in view of the numerous problems of supply of alternative fuels. Although coal is currently being imported in spite of the abundant local resources, this untenable situation is expected to change as current efforts to reactivate some local mines become a reality.

9.4.6 Nigerian coals for power generation

The plan for resuscitation of coal utilization for power generation has been discussed extensively in Chapter 9.3. Clearly, Nigerian coals have a great potential as fuel for electric power generation and current efforts to phase the resource into the national energy equation is like turning a full circle over fifty years. On the basis of technical properties, all Nigerian coals are potentially suitable for electric power generation, but, because of the relatively low carbon content and the potential of low-grade coals to retain moisture, more coal will be required per unit power output. Furthermore, boiler design in terms of size and combustion technology varies depending on coal type. The boiler required would be significantly bigger compared with bituminous coal and process equipment configuration would be different (Figure 9.11, Table 9.10).

In addition to reduced efficiency, low rank coals have lower energy density and often contain more moisture and ash, hence they require larger sized process equipment to deal with the increased mass flows. For example, brown coal (lignite) may have a moisture content as high as 60%. Also, thermal efficiency drops by around 10 percentage points compared with bituminous coal containing around 10% moisture. Low-rank coals are often dried prior to combustion but drying must be carried out immediately prior to combustion to minimize the risk of spontaneous combustion (IEA, 2011).

Four of the coal-to-power projects currently being planned in Nigeria will use sub-bituminous coals while the Gombe plant will use lignite. All indications are that the five plants will adopt clean coal technologies and this standard should be enforced in subsequent coal power plant licenses. The fact that both the coal mine and power plant are owned by the same investor and located in the same vicinity should make the projects economically competitive compared with gas-fired plants. Successful privatization of all the coal blocks and implementation of the power plant projects should enable government to concentrate on modernizing and expanding the transmission grid which is still government-owned although a foreign company has been contracted to manage the facility.

9.4.7   Production of smokeless fuel briquettes from Nigerian coals

The potential of Nigerian coals for the production of smokeless briquettes has been discussed in Chapter 7. The results of all investigations have shown that Enugu and Okaba coals have very good potential as feedstock for smokeless briquette production. About 60% of the Nigerian population still depend on fuelwood and charcoal as the primary sources of energy. Fuelwood accounts for over 50% of the total energy consumption and is the dominant source of energy in the domestic sector (NEP, 2003). It is also used in other sectors of the economy such as bakery and burnt brick cottage industries.

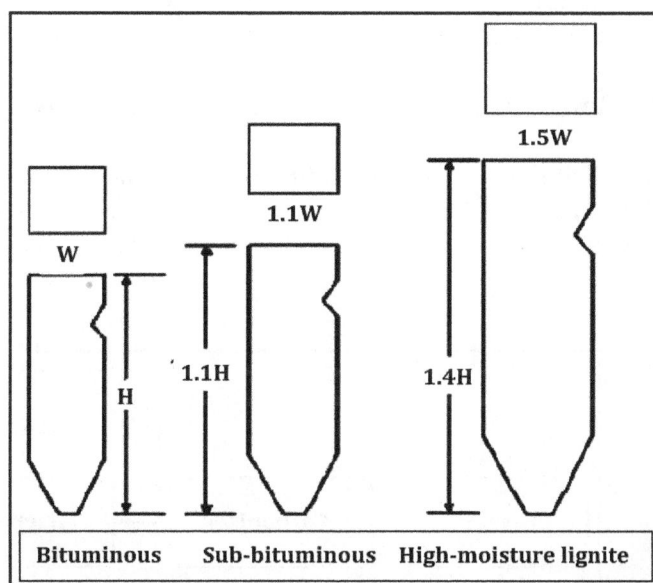

Figure 9.11   Boiler size variation with moisture content in coal.
(St Baker and Juniper, 1982).

Table 9.10 Effect of coal type on IGCC systems. *(Adapted from Korens, 2002).*

Coal type	Bituminous	Bituminous	Sub-bituminous	Lignite
Heating value (kcal/kg, ar)	7284	6116	4559	4170
Ash (dry)	7.5	12.5	17	20
Slurry concentration (% dry solids)	66	63	56	50
Relative feed rate	1	1.25	1.8	2
Number of gasifiers	2	2	3	4
Relative heat rate kcal/kWh)	1.00	1.06	1.14	1.22
Relative capital cost (per kW)	1.00	1.09	1.24	1.39

The negative implications of large-scale utilization of fuelwood and charcoal are multi-dimensional. Forests are being destroyed, leading to erosion and desertification, the hardwood resources of the country are being depleted by producers of charcoal and re-plantation strategies are virtually non-existent. Furthermore, smoke produced from the burning of fuelwood is injurious to health. Smokeless coal briquettes are in wide use in many countries including the United States, most countries in Europe, India, Russia, China, etc. and the prospects of coal-based smokeless fuel in Nigeria are very bright. The investment required is low and plants from cottage size to very large units can be operated profitably. A simple flow diagram is shown in Figure 9.12.

Production of smokeless briquettes from Nigerian coals is feasible and potentially profitable, and investment is low. It is gratifying to note that the Eta-Zuma Group that owns several coal blocks plans to process output from its lignite deposits in the Delta block into briquettes and other domestic power fuels for use mainly in the northern parts of Nigeria where charcoal is very scarce and expensive. It is hoped that this initiative will encourage entrepreneurs to consider setting up plants in areas close to the numerous coal deposits. The Ogwashi Ukwu and Maiduguri lignite deposits are potentially good sites.

9.4.8   Production of calcium carbide from Nigerian coals

Calcium carbide (CaC_2) is a valuable industrial chemical compound used mainly for the production of acetylene and calcium cyanamide ($CaCN_2$). Acetylene is used extensively in oxy-acetylene welding and calcium cyanamide is a valuable fertilizer.

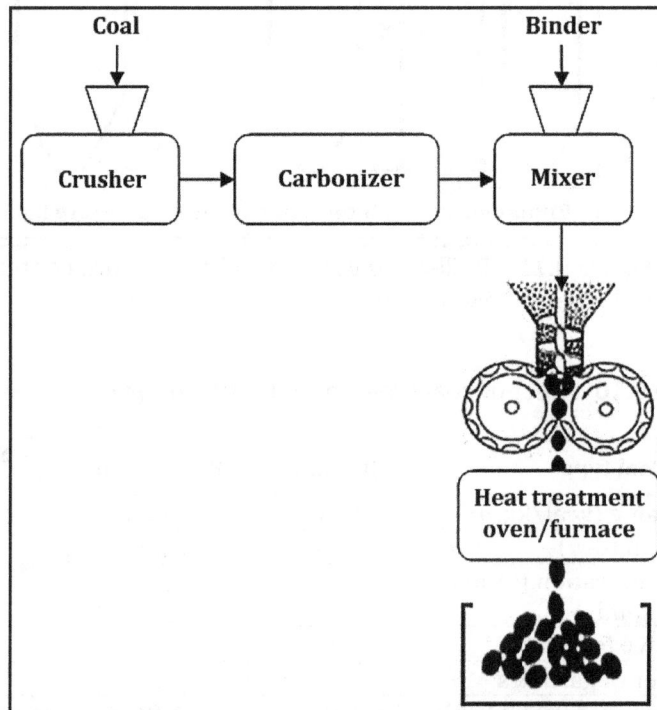

Figure 9.12   A simple flow diagram for smokeless coal production.

Calcium carbide is produced by heating a mixture of lime and coke in an electric arc furnace. Carbon reacts with lime in accordance with equation 9.1. High heat is required for the reaction and industrial production is carried out in an electric arc furnace with carbon electrodes. Calcium cyanamide is produced by heating calcium carbide in an electric furnace purged with nitrogen at around 1000°C for several hours (Equation 9.2).

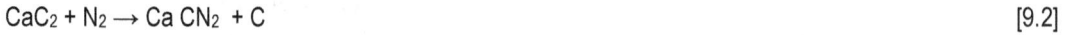

$$CaO + 3C \rightarrow CaC_2 + CO \qquad\qquad\qquad [9.1]$$

$$CaC_2 + N_2 \rightarrow Ca\,CN_2 + C \qquad\qquad\qquad [9.2]$$

The feed coke for the production of calcium carbide can be made by carbonizing any coal at high temperature of 900-1000°C. Most of the welding done in Nigeria is by the oxy-acetylene process and all the calcium carbide required is imported. The typical welder generates acetylene by throwing lumps of calcium carbide into water in a steel gas cylinder. Some companies are also producing bottled acetylene for industrial welding. All Nigerian coals are potentially suitable for the production of calcium carbide and rolling mills/foundries that already have small arc or induction melting furnaces can easily diversify into calcium carbide production.

9.4.9   Conversion of Nigerian coals to chemicals

The potential of coal as a source of hundreds of premium chemicals has been discussed extensively in Chapter 8, and available data indicate that Nigeria's high volatile coals are rich in chemicals. Only two petrochemical plants are currently operating in Nigeria. The Eleme Petrochemical plant in Rivers State, which was built adjacent to the Port Harcourt refinery in 1995, has production capacities for olefin (483,000mt/yr) , polypropylene (80,000mt/yr), and polyethylene 250,000mt/yr). Like the refinery, the plant has suffered from many technical problems, and has only operated intermittently at production levels of less than 40%. The Warri refinery in Delta State, a complex refinery which produces both gasoline and petrochemicals came on stream in 1978. The plant is managed jointly with a petrochemicals plant built in 1986 to produce 35,000mt/yr of polypropylene and 18,000mt/yr of carbon black. Again, both the gasoline and petrochemical units operate at well below capacity due to major technical and maintenance problems. A new refinery and petrochemicals plant by Dangote Investments is scheduled to commence production in 2019 in Lekki, Lagos State. The refinery which will be the largest in the world will process 660,000 barrels of oil per day. The capacity would be 150 percent of the current total demand of petroleum products in the country and the excess products would be exported to other countries. The complex also features a 1.3 million metric tonnes per annum petrochemical plant and a fertilizer plant with a capacity of 2.8 million metric tonnes of assorted fertilizers.

Coal was the main source of many major industrial chemicals until the 1960s when petroleum became the preferred precursor. The instability of the petroleum market and the relatively restricted global distribution of deposits compared with coal have shifted attention back to coal in the last decade or so. China leads the world in coal-to-chemicals and has become the world's leading producer of major chemicals and petrochemicals from coal. The fact that Nigeria has abundant gas resources may not make independent coal-to-chemicals plants economically attractive. However, hybrid plants can be designed to generate power as well as produce high premium chemicals including ethanol, methanol, ammonia, urea, olefins, dimethyl ether, ammonium nitrate and acetic acid, most of which are feedstocks for the production of other chemicals. While the level of demand of these chemicals may not justify the huge investment required, a hybrid design would introduce considerable operational flexibility. Furthermore, the potential for export of coal-based chemicals is very high. The Eta-Zuma coal-fired power plant in Kogi State has adopted a clean coal technology featuring a circulating fluidized-bed (CFB) combustor with a flue gas processing unit to recover chemicals, in particular, ammoniacal liquor and coal tar, both of which are valuable raw materials for the production of a wide range of chemicals including fertilizers. Initially, the by-products will be sold to processors but a coal tar processing line is planned for a future phase and a possible configuration is shown in Figure 9.13.

Natural gas is also an excellent feedstock for the production of premium chemicals and Nigeria has abundant reserves, rated ninth in the world. It is being exported in liquefied form and most of the country's power generating plants are running on gas. An extensive pipeline network is under construction to promote utilization by industry. However, the dynamics of the world gas market are changing and, in view of the major rise in gas production and LNG exports by several countries, in particular, United States and Russia, the market is becoming more competitive. Gas delivered through pipelines is significantly cheaper than liquefied natural gas and, with Russia already delivering gas to Europe through pipelines and planning another pipeline to China, it is expected that demand for Nigeria's gas will fall in the near future. Therefore Nigeria needs to plan ahead on aggressively expanding local demand for the resource. Gas is a cheaper and more convenient precursor to the production of chemicals than coal, hence, coal-to-chemicals projects should only be integrated with power generation. However, an aggressive development of gas-to-chemicals projects could open up a wide international market for Nigeria and help reduce the over-dependence on oil export.

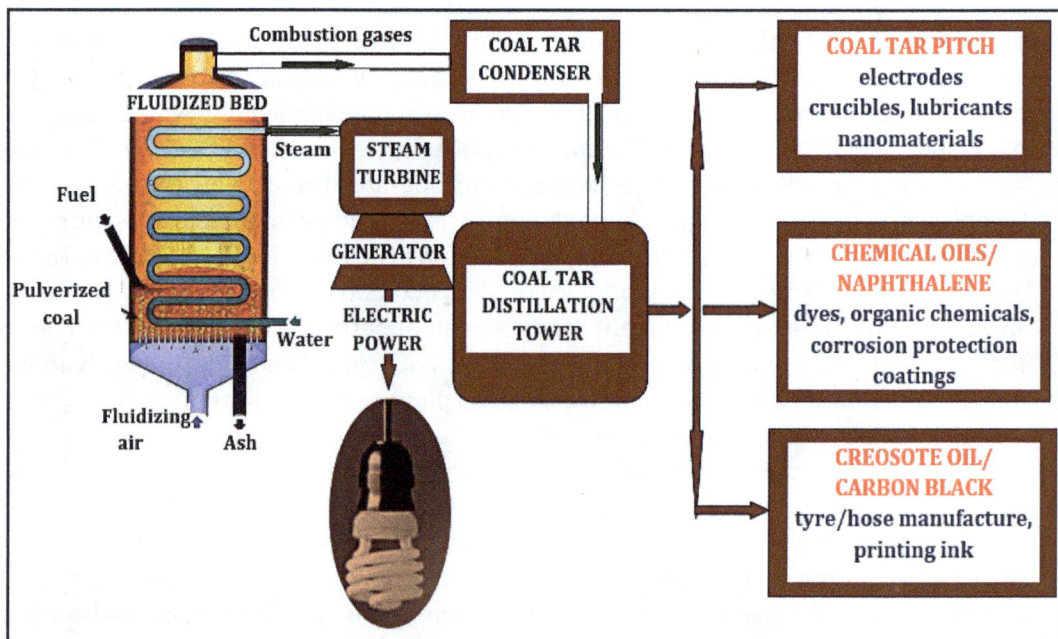

Figure 9.13   Probable configuration of the Eta-Zuma power plant at Itobe, Kogi State, Nigeria.

9.5   TECHNOLOGY OPTIONS FOR CLEAN COAL-FIRED POWER GENERATION

The classic technology for producing electric power from coal is to burn solid coal for generating steam that drives a steam turbine coupled to a generator (Figure 9.14). This technology is still in use in many developing countries. It is an inefficient technology with thermal efficiency of around 30%. A more modern technology currently in use involves pulverization of the coal before burning (Figure 9.15). Pulverization of coal prior to combustion allows for more efficient process control and higher efficiencies of around 38% are achievable. The most modern technology is to gasify the coal to produce syngas which can be used in a combined cycle (IGCC) to drive a gas turbine and a steam generator (Figure 9.16). The IGCC technology produces syngas from coal or biomass in a high pressure gasifier. The gas is then used to drive a gas turbine and waste heat is recycled for generating steam that drives a steam turbine.

The production of syngas also makes it possible to configure a power plant with a coal-to-chemicals unit. The IGCC technology is not only the most efficient (around 40% thermal efficiency), flue gases can be captured and processed more easily because the gas is cleaned before the gas turbines, when it is at high pressure and more concentrated, In contrast,flue gases are cleaned after combustion when the exhaust is diluted and at low pressure. Emissions of particulate matter, SOx, NOx are significantly lower and carbon dioxide capture is also significantly easier when the gas is at high pressure (Figure 9.17). Furthermore, the syngas can be used as feedstock for an auxiliary chemicals plant. However, capital and operating costs compared with pulverized coal power generation are 20-30% higher.

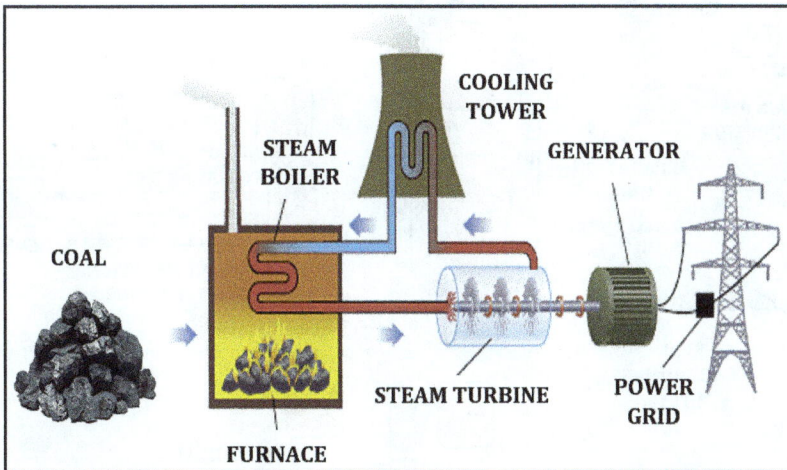

Figure 9.14   Options for utilization of Nigerian coals for electric power generation: Solid coal-fired power generating plant. *(revisionworld.com).*

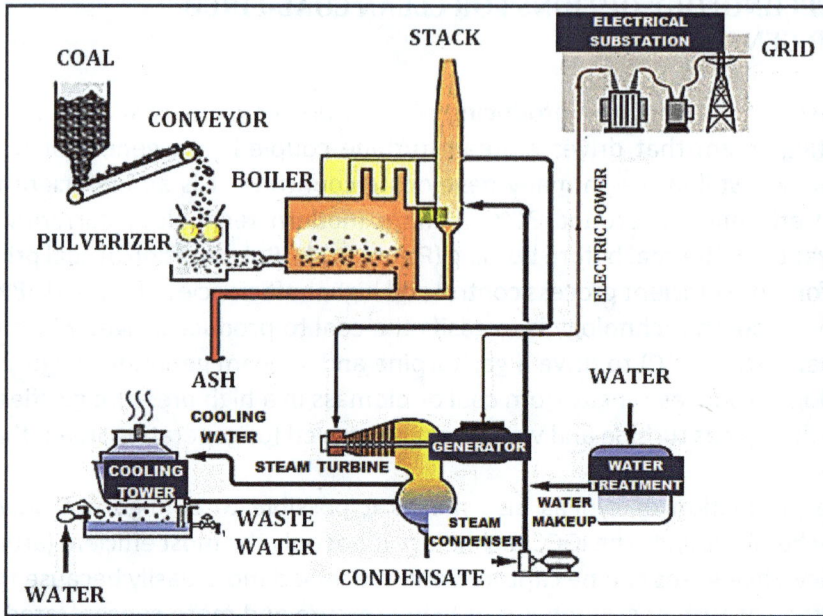

Figure 9.15 Options for utilization of Nigerian coals for electric power generation: Pulverized coal-fired power generating plant. *(blog.gerbilnov.com)*.

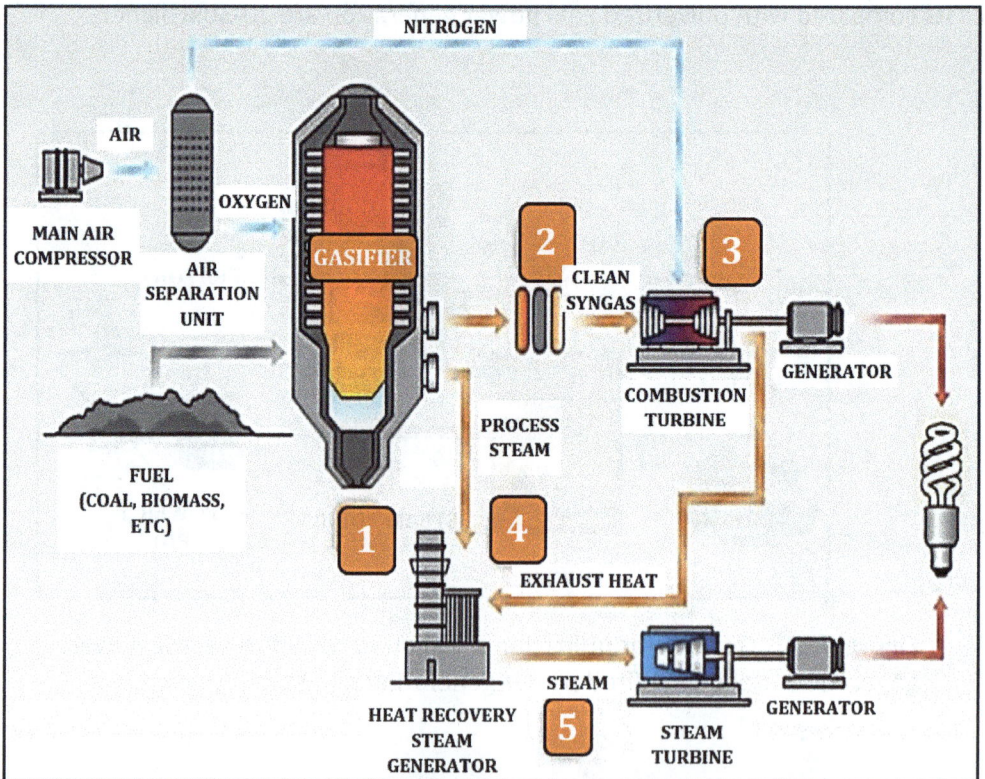

Figure 9.16 Options for utilization of Nigerian coals for electric power generation: Integrated combined cycle coal-to-syngas power generating plant. *(Dukeenergy.com)*.

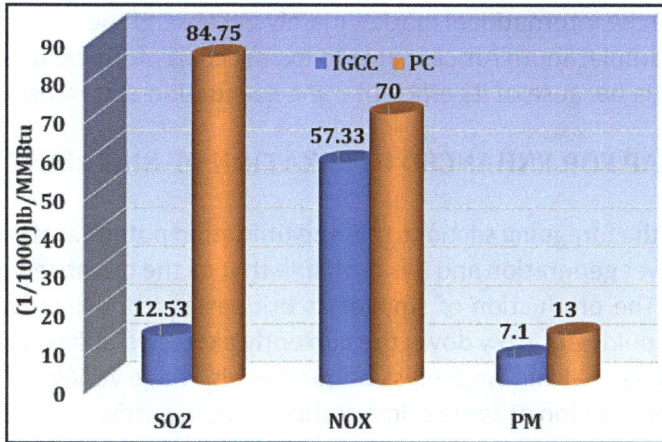

Figure 9.17 Average emissions comparison for sulphur dioxide (SO$_2$), nitrogen oxides (NOX) and particulate matter (PM) between IGCC and pulverized coal. (PC super- and sub-critical) power plants, without carbon capture *(DOE/NETL, 2010).*

9.6 EXPORT POTENTIALS OF NIGERIAN COALS

Many publications stress the high potential of Nigerian coals for export. However, available data on the technical quality of all proven reserves of Nigerian coals do not support this option. The bulk of world trade in coal involves coking coals and hard, bituminous coals. One of the main parameters for assessing the export value of coal is the heat content per unit cost of transportation. On this count, Low grade sub-bituminous coals and lignites are of low value. Another variable is the sulphur content which is low in most Nigerian coals hence this is a positive asset. A third important variable in assessing coal for export is its propensity for weathering. Coal for export is usually stocked in piles at mines, in ports and on ships for weeks before delivery to the customer, and also in customers' stockyards. The heat value of coal that weathers easily could deteriorate significantly within weeks. Furthermore, oxidative reactions which take place in moist coal piles could generate enough heat to self-ignite the coal pile. Nigerian coals weather significantly and lose heat value in storage. Also , incidents of self-ignition have occurred during storage in mines, in customers' stockyards, and on ships carrying coal to some neighbouring countries. These incidents prompted the Nigerian Coal Corporation to commission an extensive investigation of the weathering characteristics of Enugu and Okaba coals which were the only deposits being mined commercially at the time (Afonja, 1973).

Most of Nigeria's coal deposits are not close to a seaport or linked by a rail line hence exportation even to neighbouring African countries would be problematic. The exception is Enugu coal deposit which can be transported by rail to Port Harcourt and shipped. Also, the coal is the most stable and can be exported to countries on the west coast of Africa. Most countries that import coal hold several months' requirement in storage in case of delays in delivery. Also, importers in temperate countries accumulate large stocks in summer to cope with the high demand in winter. Apart from providing elaborate storage facilities, customers also prefer coals that are stable over long periods of storage. It is unlikely that Nigerian coals can compete in the international market mainly because of the poor internal transportation system. For example, South Africa is able to export coals of similar quality to other countries including Nigeria because of its efficient mine evacuation and transportation logistics.

9.7   ROADMAP FOR ENHANCED UTILIZATION OF NIGERIAN COALS

It is clear from the foregoing sections that the utilization potential of Nigerian coals is highest for electric power generation and it is probable that all the privatized deposits will feed local power plants. The production of smokeless briquettes from Nigerian coals also has great potential and could help slow down the currently extensive utilization of the country's hard wood for charcoal making. Smokeless briquettes also have very good export potential. The prospects for production of some primary chemicals, in particular, calcium carbide,benzene, methanol are also very good. Chemicals' production units can be integrated with power production, partly to capture anthropogenic by-products of coal combustion, but also to produce valuable chemicals for local consumption and export. For example most neighbouring countries also have a problem with deforestation due to processing of hardwood into charcoal and would be good potential markets for smokeless coal briquettes. A feasible roadmap for maximal utilization of Nigerian coals is presented in Figure 9.18.

9.8   IMPLICATIONS OF THE PRIVATIZATION
OF NIGERIAN COAL DEPOSITS

The decision by the Nigerian government to privatize the Nigerian coal deposits and the numerous other non-performing national industrial projects was one of the most important in the last decade or so. It put an end to government interference and gave the coal industry a new lease of life. Previous attempts to revitalize the Nigerian coal Corporation through partnership with foreign and local companies failed but, with this move, around eighty foreign companies and foreign-local partners have submitted bids for coal mines and licenses to prospect for coal.

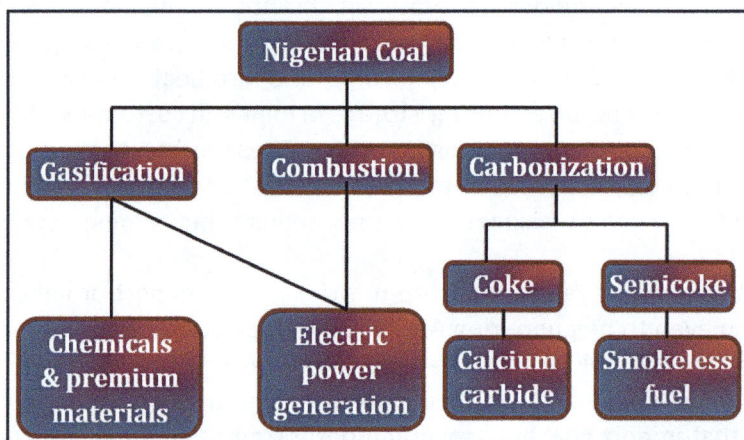

Figure 9.18 Roadmap for maximal utilization of Nigerian coals.

The implication of privatization is that government no longer has direct control of the industry and, from past experience, companies that win bids tend to deviate from terms of the privatization agreement. The problems that have emerged from the privatization of the automobile and steel industries are typical examples. It is important therefore that each coal mining concession is tied to power generation and an effective monitoring unit is in place to regulate the activities of new owners of privatized coal mines to ensure that they do not simply mine coal and export to some neighbouring countries. Another Important area for strict monitoring is compliance with set environmental standards.

REFERENCES

Afonja, (1973). "An investigation of the spontaneous combustion propensity of Nigerian Coals." Research Report: IFE/CRL/NCC/7. Nigerian Coal Corporation.

Afonja, A. A. (1979). "The Nigerian Iron and Steel Project: Problems and Prospects." *Proc. of the Symposium of the Nigerian Society of Chemical Engineers,* Ile-Ife, 21 – 28.

Afonja, A. A., and B. A. Ajakaiye (1981). "Problems and Prospects of Steel Development in Nigeria: A Study of the Interface Problems of Five National Steel Projects." A report submitted by MaMiCh Consultancy Unit, Faculty of Technology, University of Ife to Steel Development Department, Executive Office of the President, Federal Republic of Nigeria.

Burnard, K., and S. Bhattacharya, S. (2011). "Power generation from coal: Ongoing developments and outlook." OECD/IEA, 2011.

DOE/NETL (2010) "Cost and performance Baseline for Fossil Energy Plants, Vol. 1, DOE/NETL-2010/1397.

EIA (2016). *International Energy Outlook 2016.* U.S. Energy Information Administration. eia.gov.

Eta-zuma, I (2014). "Coal to Power Project in Kogi." Presentation at the meeting with Minister of Power. kogireports.com; etazuma.com.

IEA (2016). *Key World Energy Statistics 2016.* International Energy Agency, www.iea.org.

Korens, N. SFA Pacific, "Process screening analysis of alternative gas treating and sulfur removal for gasification", Prepared for DOE/NETL, 2002.

Lukman,R.(2003). *National Energy Policy.* Energy Commission of Nigeria, April, 2003).

Nebo, C. (2015). "How we plan to diversity electricity generation using coal, solar." Minister of Power. The Guardian, January 04, 2015.

NEP (2003). *National Energy Policy.* Energy Commission of Nigeria www.energy.gov.ng.

OECD/IEA (2012). *World Energy Outlook.*

Oyeyinka-Oyelaran, B. and O. Adeloye (1995). "Technological change and project execution In Nigeria: The case of Ajaokuta Steel Plant." Technology Policy and Practice in Nigeria." Ogbu, O. Oyeyinka, O. and M. Miawa (Eds). In *Technology Policy and Practice in Africa.* International Development Research Centre (IDRC), Ottawa, Canada.

Roadmap (2010). "Roadmap for Power Sector Reform." Presidential Task Force on Power, Nigeria.

Roadmap (2013). "Roadmap for Power Sector Reform. Revision 1" Presidential Task Force on Power, Nigeria.

CHEMICALS FROM COAL

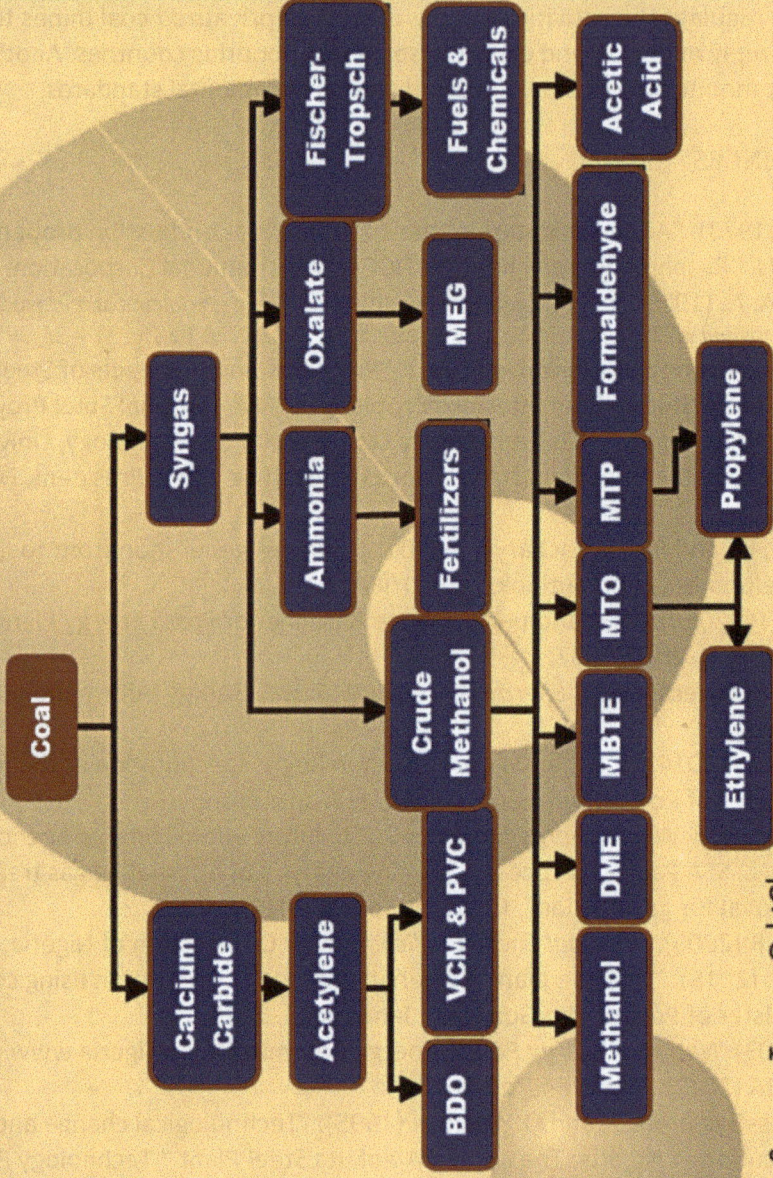

Source: Tecnon OrbiChem

Index

41% OF GLOBAL ELECTRICITY COMES FROM COAL
(Image from American Coalition for Clean Coal Electricity)